Springer Theses

Recognizing Outstanding Ph.D. Research

Aims and Scope

The series "Springer Theses" brings together a selection of the very best Ph.D. theses from around the world and across the physical sciences. Nominated and endorsed by two recognized specialists, each published volume has been selected for its scientific excellence and the high impact of its contents for the pertinent field of research. For greater accessibility to non-specialists, the published versions include an extended introduction, as well as a foreword by the student's supervisor explaining the special relevance of the work for the field. As a whole, the series will provide a valuable resource both for newcomers to the research fields described, and for other scientists seeking detailed background information on special questions. Finally, it provides an accredited documentation of the valuable contributions made by today's younger generation of scientists.

Theses are accepted into the series by invited nomination only and must fulfill all of the following criteria

- They must be written in good English.
- The topic should fall within the confines of Chemistry, Physics, Earth Sciences, Engineering and related interdisciplinary fields such as Materials, Nanoscience, Chemical Engineering, Complex Systems and Biophysics.
- The work reported in the thesis must represent a significant scientific advance.
- If the thesis includes previously published material, permission to reproduce this must be gained from the respective copyright holder.
- They must have been examined and passed during the 12 months prior to nomination.
- Each thesis should include a foreword by the supervisor outlining the significance of its content.
- The theses should have a clearly defined structure including an introduction accessible to scientists not expert in that particular field.

More information about this series at http://www.springer.com/series/8790

Maohui Luo

The Dynamics and Mechanism of Human Thermal Adaptation in Building Environment

A Glimpse to Adaptive Thermal Comfort in Buildings

Doctoral Thesis accepted by
Tsinghua University, Beijing, China

清華大學出版社
TSINGHUA UNIVERSITY PRESS

Springer

Author
Maohui Luo
Tongji University
Shanghai, China

Supervisors
Prof. Yingxin Zhu
Department of Building Science
Tsinghua University
Beijing, China

Associate Prof. Bin Cao
Department of Building Science
Tsinghua University
Beijing, China

ISSN 2190-5053 ISSN 2190-5061 (electronic)
Springer Theses
ISBN 978-981-15-1164-6 ISBN 978-981-15-1165-3 (eBook)
https://doi.org/10.1007/978-981-15-1165-3

Jointly published with Tsinghua University Press
The print edition is not for sale in China (Mainland). Customers from China (Mainland) please order the print book from: Tsinghua University Press.

This Springer imprint is published by the registered company Springer Nature Singapore Pte Ltd.
The registered company address is: 152 Beach Road, #21-01/04 Gateway East, Singapore 189721, Singapore

Supervisor's Foreword

Adaptive thermal comfort in the built environment is a complex problem involving anthropology, physiology, psychology, architecture, and heat transfer disciplines. It is also a frontier problem in the thermal comfort research field and of great practical value in sustainable building environment conditioning technologies.

This thesis studied how people with different thermal experiences adapt to different indoor climates and the underlying reasons behind the adaptation. A series of thermal comfort surveys and climate chamber experiments were conducted to understand building occupants thermal adaptation from perspectives of physical parameters, physiological acclimation, psychological adaptation, and model development. The following work and findings are noteworthy.

It studied the dynamic characteristics of human thermal adaptation in buildings. It was found that people's understandings of thermal comfort are mutually dependent on their indoor thermal experiences. Also, building occupants' thermal adaptation exhibits asymmetric trajectories: It is much quicker for occupants to raise their expectations and accept a neutral indoor climate than to lower their expectations and acclimate to under-conditioned environments.

It also investigated the basic laws of physiological acclimatization, and psychological adaptation and their effects. Based on the timescale, physiological adaptation was divided into long-term acclimatization and short-term regulation. It was found that the comfort improvement by environmental control approach comes from psychological factors of perception control and physical factors of environmental changes, of which the former is more obvious in the environment deviating from thermal neutrality.

Based on the classical heat transfer model of PMV and the theoretical framework of thermal adaptation, an adaptive heat balance comfort model has been developed, which has guiding significance for the development of future adaptive comfort models.

Beijing, China
July 2019

Prof. Yingxing Zhu

Abstract

Adaptive thermal comfort research aims to achieve a balance between energy saving in building environment conditioning and comfort improving in occupants' thermal perception. This thesis focused on how people with different thermal experiences adapt themselves to various indoor climates and the explanations behind such adaptation. Both filed investigation in real buildings and human subject experiments in climate chamber were conducted to reveal the basic laws underpinning building occupants' thermal adaptation from perspectives of physical improvement, physiological acclimation, psychological adaptation, and model development. The following work and findings are noteworthy.

First, online surveys were carried out to explore the thermal adaptation phenomenon among Chinese residents with different indoor climate experience. The results showed a considerable mutually dependent relationship between residents' thermal comfort expectation (or demand) and indoor climate experience. People with comfortable thermal experience tended to be pickier on their current thermal environments while those experienced lower-grade thermal conditions were more likely to forgive the current thermal uncomfortable.

Second, comparative thermal comfort investigations were conducted in real buildings in Shanghai and Beijing. The results showed that building occupants' thermal adaptation exhibits asymmetric trajectories. It is much quicker for occupants to raise their expectations and accept a neutral indoor climate than to lower their expectations and acclimate to under-conditioned environments. The demand factor was proposed to describe such asymmetric adapting processes.

Third, physiological experiments were conducted to examine how physiological acclimation affects subjects' thermal comfort. The results showed that physiological acclimation can be separated into two categories: short-term regulation and long-term acclimation. The short-term physiological regulation can be reflected by metabolic rate changes while the long-term acclimation described by indexes like weighting factor based on skin temperature and blood flow differences.

Fourth, psychological experiments were conducted in a climate chamber to explore how personal control of the indoor environment influence subjects' thermal comfort psychologically. It showed that the perceived control improved occupants' thermal comfort in unneutral thermal conditions and the effects can be quantified by linear regressions.

Lastly, the adaptive heat balanced comfort model was established based on a framework including different adaptive layers such as behavioral adjustment, physiological acclimatization, psychological adaptation, and physical improvement. The new model inherited the computational advantages of PMV and the empirical advantages of adaptive comfort model, which makes it more flexible in practical applications.

Keywords Adaptive thermal comfort · Thermal comfort expectation · Physiological acclimation · Psychological adaptation

Acknowledgements

This study was supported by the China National Key R&D Program (No. 2016YFC0700100), and the National Natural Science Foundation of China (No. 51521005 and No. 51678330). The project was performed within the framework of the International Energy Agency—Energy in Buildings and Communities Program (IEA-EBC) Annex69 "Strategy and Practice of Adaptive Thermal Comfort in Low Energy Buildings".

Contents

Nomenclature

Abbreviations

AC	Air-conditioning
AHBM	Adaptive heat balance model
APMV	Adaptive predicted mean vote
BAT	Brown adipose tissue
BMI	Body mass index
ET	Effective temperature
HVAC	Heating, ventilation, and air-conditioning
kgce	Kilogram standard coal equivalent
MM	Mixed-mode
NV	Nature ventilation
PCS	Personal comfort system
PMV	Predicted mean vote
PPD	Predicted percent dissatisfied
SET	Standard effective temperature
tce	Ton standard coal equivalent
TCV	Thermal comfort vote
TSV	Thermal sensation vote
UTCI	Universal thermal climate index
WBGT	Wet-bulb globe temperature

Physical Variables

λ	Adaptive coefficient
v	Airspeed (m/s)
t_a	Air temperature (°C)
t_g	Black globe temperature (°C)

Q_{CO_2}	Carbon dioxide production flow rate (ml/s)
I	Clothing insulation (clo)
t_{cl}	Clothing outer surface temperature (°C)
t_{comf}	Comfortable indoor temperature (°C)
h_c	Convective heat transfer coefficient (W/(m^2·K))
A_d	Dubois body surface area (m^2)
f_{cl}	External body surface area ratio to bare body surface area
H	Height (m)
M	Human body metabolic rate (W/m^2)
T_{out}	Mean outdoor air temperature (°C)
t_r	Mean radiant temperature (°C)
t_{ms}	Mean skin temperature (°C)
W	Mechanical work of human body (W/m^2)
ΔM	Metabolic rate change (met)
$t_{out,m}$	Monthly mean outdoor temperature (°C)
t_{op}	Operative temperature (°C)
Q_{O_2}	Oxygen consumption flow rate (ml/s)
RH	Relative humidity (%)
RQ	Respiratory coefficient
A_{ch}	Seat and body contact area (m^2)
e	Thermal expectation factor
I_{ch}	Thermal resistance of the seat (clo)
P_a	Water vapor pressure in air (Pa)
WE	Weight (kg)

Chapter 1
Introduction

1.1 Background

China building sector has been growing rapidly in recent years. From 2000 to 2013, the urban residential area increased from 9.5 to 23.4 billion m^2. Figure 1.1 shows that Chinese total building energy consumption increased from 320 to 756 million tce. Under this context, how to effectively control building energy consumption and improve building environment quality has become a hot topic. According to a strategy named 'total energy consumption control', if the total building area in the country will be 80 billion m^2 in future, the corresponding energy consumption in should be controlled at 1.1 billion tce. Which means the maximum building energy intensity is

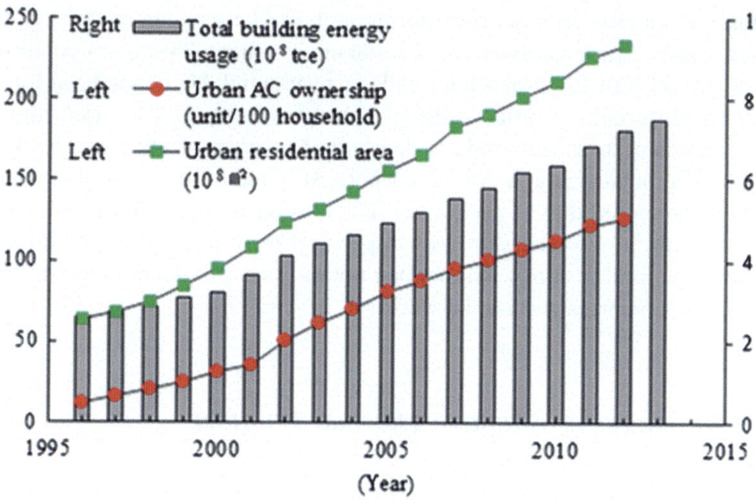

Fig. 1.1 China's urbanization process from 1996 to 2013 [2]

M. Luo, *The Dynamics and Mechanism of Human Thermal Adaptation in Building Environment*, Springer Theses,
https://doi.org/10.1007/978-981-15-1165-3_1

13.7 kgce/m^2, equivalent to 30% of current building energy intensity in the United States or 40% of that in Japan and Germany [1].

With the development of economy and living standard improvement, Chinese people's pursuit of building environment quality is getting higher, caring more about comfort and health aspects of buildings. It is essential to create sustainable indoor environment without exceeding the upper energy intensity limit. Among all the building energy consumption ends, heating, ventilation, and air conditioning (HVAC) system take up 40% of the total building energy consumption [1]. Proper design and operation of the HVAC system can help to meet occupants' health and comfort requirement meanwhile lower the energy expenditure.

1.1.1 Practical Values of Thermal Comfort Research

Thermal comfort studies, especially adaptive thermal comfort research, can help to fulfill this aim by providing a basis for indoor thermal environment evaluation. Firstly, the adaptive thermal comfort phenomenon can support the climatic building design strategy other. The traditional architecture forms in different climate zones are a historical confirmation of how buildings should adapt to local climate [3–5]. North China is hot humid in summer but cold and dry in winter. So that its local 'Si He Yuan' buildings (Fig. 1.2a) can keep residents warm in winter but provide shading in summer. The "Cave dwelling" in the loess plateau in northwest China (Fig. 1.2b) is well adapted to climates with large annual temperature variance. In southwest China, residents created the "Gan Lan" building (Fig. 1.2), which successfully protect occupants from rain, moisture, and heat.

In addition, adaptive thermal comfort research can help understand the changing trend of people's requirements on the thermal environment. Taking air conditioning as an example. Figure 1.3 shows the trends of residential AC penetration (or adoption rates) in different countries. Although the growing periods varied, the sharp increasing trends in these economic or demographic superpowers were shocking. A study [6] looked in particular at Mexico, where AC penetration was only about 13% in 2012 and forecasted that as people get richer, those living in warm climates will flock to AC with 2.7% ownership growth per $1000 of annual household income. More worrying, similar things will happen not only in Mexico but also in low and middle-income countries around the world.

1.1.2 Challenges in Adaptive Thermal Comfort

In the long history of mankind, the use of fire extended the range of human activity from tropical areas to colder areas. It has been widely used in buildings as fire basin, fire Kang, firewall, fireplace, etc. Among which the fireplace is still in use in modern

(a) Si HeYuan (b) Cave dwelling

(c) Anhui architecture (d) Gan Lan

Fig. 1.2 Typical climate adaptive buildings in China

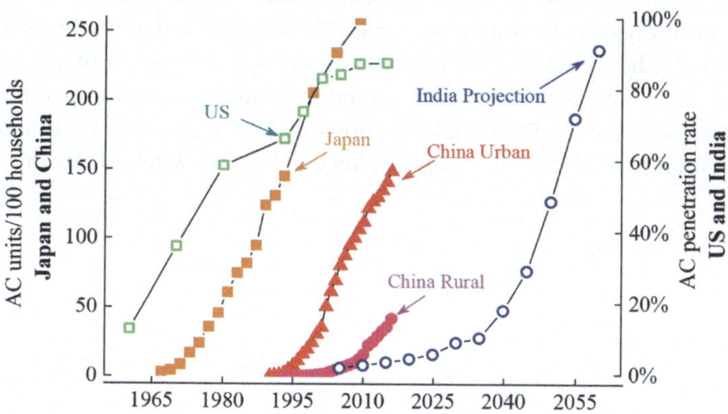

Fig. 1.3 Increasing trend of air conditioner penetration or adoption rate (US data is from Ref. [7], Japanese data is from Ref. [8], Chinese data is from Ref. [2], India data is from Ref. [9])

buildings. Compared to fire, mechanical heating and cooling like HVAC devices have a much shorter history. The first electric air conditioner was invented until 1902 by Dr. Carrier [10]. Early air conditioning aimed to meet manufacturing process demand.

Figure 1.4 shows a brief history of thermal comfort research. In the 1970s, due to the popularization of HVAC products, intensive thermal comfort studies were conducted to answer how do different thermal conditions influence human thermal comfort. Professor Fanger in Denmark Technology University developed the Predicted Mean Vote—Predicted Percentage of Dissatisfaction (PMV-PPD) model [11] based on human body heat balance to correlate subjective comfort status with objective thermal conditions. Since then, the model has been widely adopted by standards such as ISO 7730:2005 [12], American standard ASHRAE Standard 55-2013 [13], European standard CEN 15251 [14] and Chinese national standard GB/T 50785 [15].

Different from the PMV model, adaptive thermal comfort research pays more attention to building occupants' role in maintaining themselves in comfort status [16], and the adaptive models based on field studies are closer to occupants' actual feeling [17–20]. Among the existing adaptive models, the one developed by Professor de Dear [21] is the most widely influenced.

However, the PMV and adaptive comfort model are two separate models and they may be incompatible sometimes. When applying these two models in real buildings, there may occur awkward issues. For example, the current standards simply classify buildings into two categories, i.e., air conditioning (AC) buildings and nature ventilation (NV) buildings, in a black-and-white way. The PMV model is suitable for AC buildings, while the adaptive model is suitable for NV buildings. But in the real world, buildings can be much more complicated. Given that mixed-mode (MM) buildings usually have mechanical air conditioning that coexists with natural passive cooling [22, 23], there is no consensus on the thermal comfort standard that should be applied to mixed-mode buildings. As shown in Fig. 1.5, if different models were applied to evaluate MM buildings, the results can be completely different.

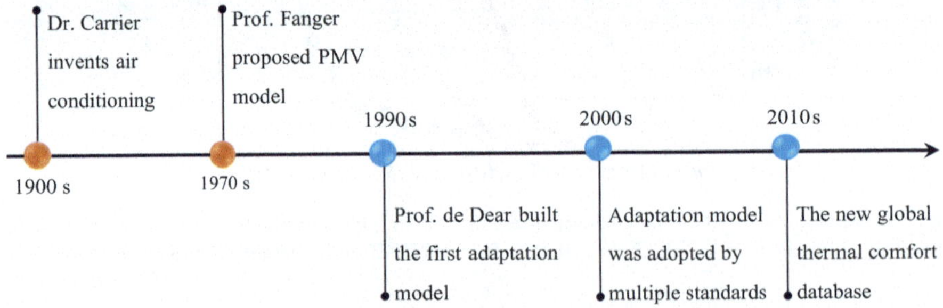

Fig. 1.4 A brief summary of thermal comfort research

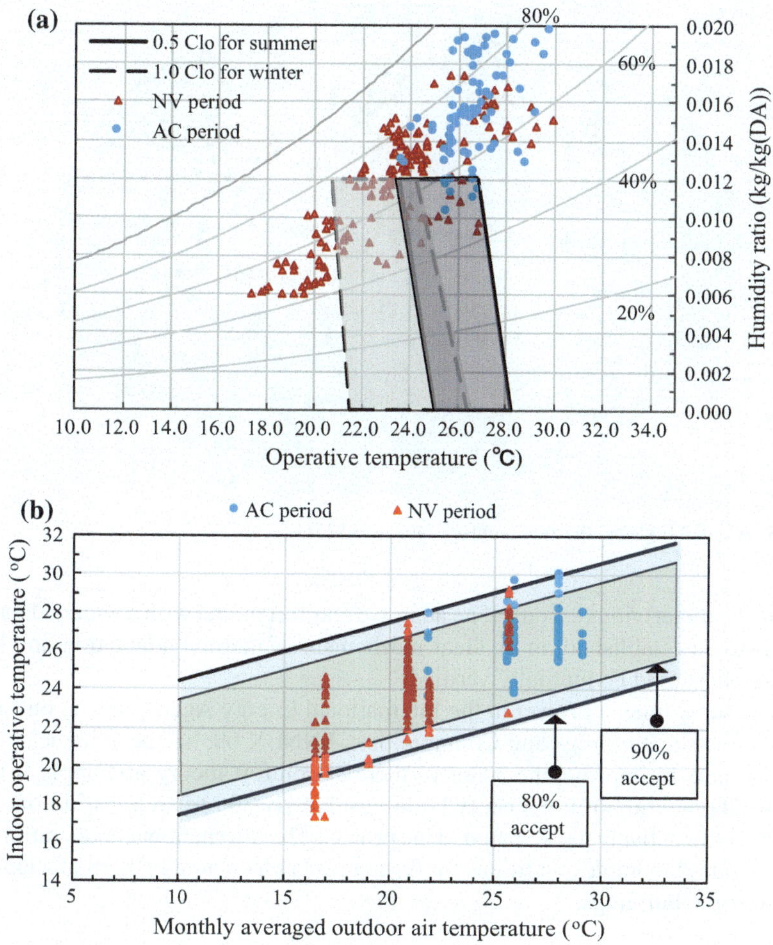

Fig. 1.5 Thermal comfort in MM building: **a** PMV model, **b** adaptive comfort model

1.1.3 New Progress in Adaptive Thermal Comfort Research

Although adaptive thermal comfort research already has a nearly 30-year history, it is still a hot topic in the field. In the past several years, a new thermal comfort database for building environment has been gradually established, including 6 continents, 23 countries, and 81,846 samples (see Fig. 1.6). The new database aims to make up the defects in the first ASHRAE thermal comfort database (RP-884) in terms of data uniformity, sample representativeness and time validity [24]. So that it is promising to establish adaptive thermal comfort model with a wider application range. However, large-scale empirical data can only reflect the phenomenon while

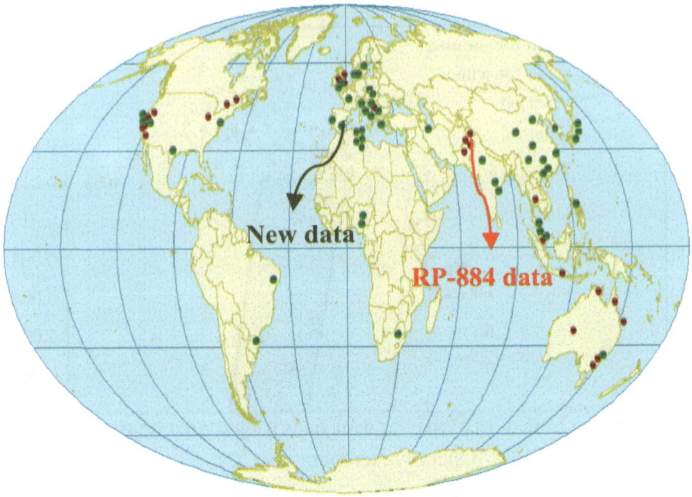

Fig. 1.6 ASHRAE global thermal comfort databases [25]

missing the underlying reasons. If an adaptive comfort model with a wider adaptation range is to be established, more clear mechanism of thermal adaptation and larger data sample should be mutually verified.

At the same time, a project in the International Energy Agency (IEA) on Energy Conservation in Buildings and Communities, ANNEX 69, has been launched. The proposal aims to improve the adaptive thermal comfort theory and apply it to the design and operation of low-energy buildings. Nearly 30 research institutions in 12 countries have actively participated in the project. The international cooperation platform provides favorable conditions for further research on human thermal adaptation mechanism in buildings.

1.2 Literature Review

Generally, thermal comfort research involves many specific areas, such as outdoor thermal comfort and urban planning [26, 27], indoor thermal comfort and productivity [28–30], thermal comfort and building energy consumption [31, 32], thermal comfort under different scenarios [33–37] and so on. Given the objectives of this thesis, the above topics will not be introduced in detail here. Relevant contents can be referred to relevant reviews [38, 39]. This section mainly summarizes steady-state thermal comfort studies and adaptive thermal comfort studies.

1.2.1 Thermal Comfort Introduction

Thermal comfort study in buildings belongs to physiological psychology research. It involves physics, physiology, psychology, and even culture disciplines. Figure 1.7 shows the relationships between building environment and human body heat transfer. The building envelope and clothing together form a microclimate that affects our subjective feelings meanwhile involving different forms of heat and moisture transfer. Human body temperature regulation involves multiple physiological regulation processes. Human perception of the surrounding environment involves psychological processes. All these multi-discipline interactions increase challenges to thermal comfort research. Under such background, the basic method of thermal comfort research is objective quantification of subjective feelings, i.e. using a series of quantification rules to correlate subjective feelings with physical heat balance status.

'Thermal sensation' is a commonly used term in thermal comfort research to describe the subjective feeling of the human body after being stimulated by cold and heat. Since subjective perception cannot be directly measured by instruments, questionnaires like thermal sensation vote (TSV) are used to quantify the 'sensation'. Historically, there have been different TSV scales [41, 42], of which the 7-point scale (shown in Fig. 1.8) recommended by ISO [12] and ASHRAE standard 55 [13] is the most widely used. TSV can effectively correlate subjective thermal sensation and objective thermal condition.

Similarly, to quantitatively describe the subjective 'comfort' perception, thermal comfort vote (TCV) is commonly used. Figure 1.9 shows a quantitative voting scale

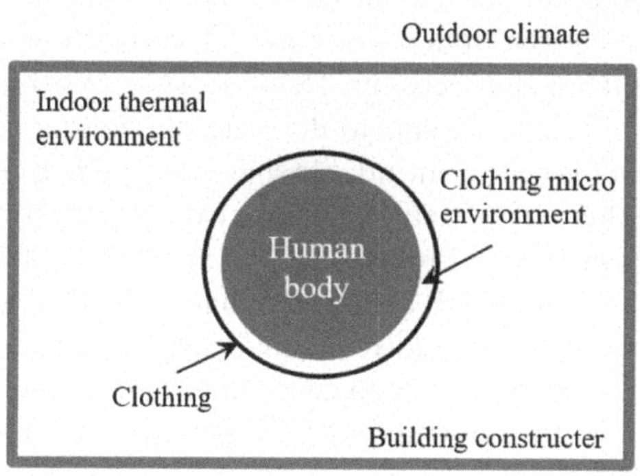

Fig. 1.7 Buildings, occupant, clothing, and the environment [40]

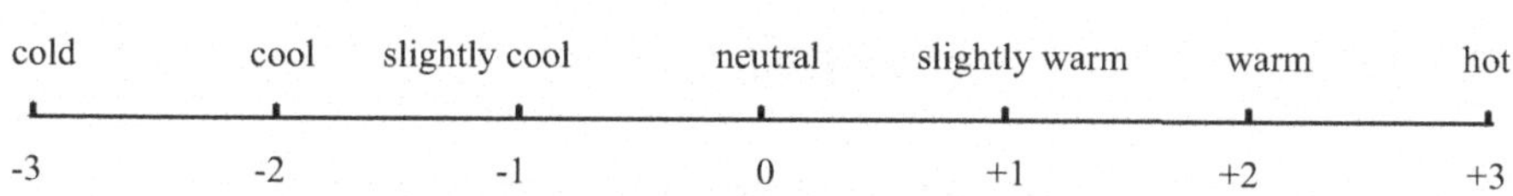

Fig. 1.8 Seven-point scale TSV

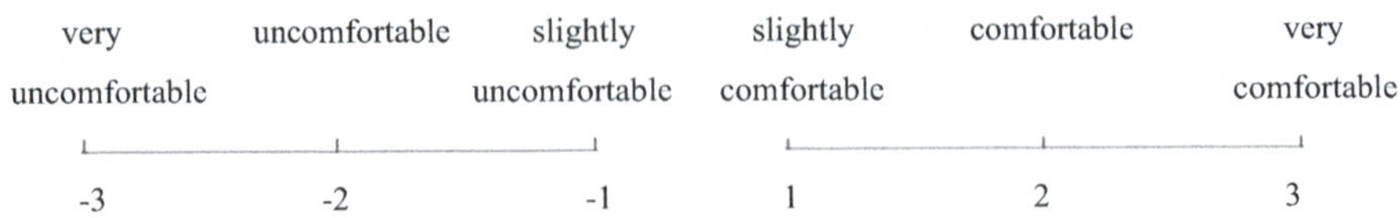

Fig. 1.9 Thermal comfort vote scale

Fig. 1.10 The correlations between TSV and other comfort votes [46–49]

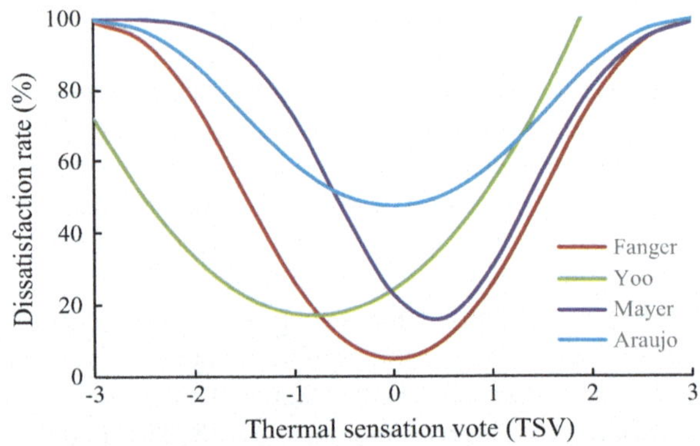

of TCV. In addition, other comfort metrics like Satisfaction Rate Vote, Thermal Preference Vote, Humidity Perception Vote, Air Movement Preference Vote are widely used as well [43].

There is a certain correlation between TSV and other comfort voting. So that TSV is often used as an indicator to reflect comfort and satisfaction status. For example, Fig. 1.10 shows the relationship between TSV and TCV. When thermal sensation is neutral (i.e. TSV $= 0$), it corresponds to the state of "neither cold nor hot". Many researchers believe that the most comfortable state occurs when our body has the least thermal stress. When thermal sensation deviates from neutrality (i.e. TSV $\neq 0$), the heat balance status of our body is broken, thus the thermal comfort perception would decline correspondingly [11, 44]. It is believed that when TSV is maintained within ± 0.5, over 90% of occupants should feel comfortable and satisfied. When TSV is maintained within ± 1, over 80% of occupants should feel comfortable and satisfied. According to this assumption, maintaining TSV between -1 and 1 is widely used in real HVAC control [45].

1.2.2 Heat Balance Comfort Models

1.2.2.1 Comfort Indexes Related to Human Body Heat Balance

Factors that affect human body heat balance include air temperature, radiation, relative humidity, air movement, clothing insulation, activity intensity, etc. [50]. Heat balance-related model describes the influence of different thermal parameters on human body heat dissipation according to heat transfer principles. And then correlate human body heat balance state to subjective thermal sensation. To date, several comfort indexes have been put forward. Table 1.1 summarizes the major features of these indexes.

The predicted mean voting (PMV) considers environmental parameters (air temperature, air velocity, radiant temperature, and relative humidity) and personal factors (activity intensity and clothing insulation) as input variables, and then calculates heat balance status of the human body and associates it with the seven-point TSV. The operative temperature (t_{op}) comprehensively considers the influence of air temperature and radiant temperature on human body heat balance. ET, SET, and WBGT can convert other thermal parameters like solar radiation into the influence of air temperature. UTCI calculates skin temperature through a multi-node heat dissipation model of the human body and then correlates TSV with skin temperature.

Among the indexes listed in Table 1.1, PMV and SET are adopted by the ISO 7730:2005 standard, American ASHRAE Standard 55-2013, European CEN 15,251 standard, and Chinese GB/T 50785 standard, and etc. The PMV is widely used for normal building environment evaluation while the SET is mainly for air movement evaluation. Considering the limited space and the need for the coming sections, we mainly emphasize the PMV equations and their derivations. For other indexes, please refer to relevant literature.

PMV uses six parameters (e.g. air temperature t_a, mean radiant temperature t_r, relative humidity RH, air velocity v, clothing insulation I, and metabolic rate m) to establish the human body heat balance equation (see Eq. 1.1). Then, it correlates human body heat load with subjective thermal sensations according to climate chamber human subjects test results. A detailed derivation of PMV can be found in reference [11, 46]. Its calculating code can be found in ASHRAE Standard 55, GB/T 50785 and other standards. A web site to calculate PMV can be found in [57].

Table 1.1 Comparison of thermal environment evaluation indexes

Indexes	Input parameters	Major features
PMV [11]	I, M, t_a, t_r, RH, v	Calculate human body heat balance state, suitable for the steady-state neutral thermal environment
t_{op} [51]	t_a, t_r	Consider radiation temperature and air temperature
SET [52–54]	t_a, t_r, RH, v, I, M	Calculate human body heat balance with consideration of air movement based on the two-node model
UTCI [55]	t_a, t_r, RH, v, I, M	Calculate human body heat balance based on the multi-node physiological thermal regulation model
WBGT [56]	t_a, t_r, RH, v	Suitable for outdoor thermal comfort evaluation with consideration of solar radiation and wind speed

$$PMV = \left(0.303e^{-0.036M} + 0.0275\right)\{M - W - 3.05[5.733 - 0.007(M - W) - P_a] \\ - 0.42(M - W - 58.2) - 0.0173M(5.867 - P_a) - 0.0014M(34 - t_a) \\ - 3.96 \times 10^{-8} f_{cl}\left[(t_{cl} + 273)^4 - (t_r + 273)^4\right] - f_{cl}h_c(t_{cl} - t_a)\} \quad (1.1)$$

In the equation, P_a is partial pressure of water vapor in air (Pa), t_a is ambient air temperature (°C), f_{cl} is the ratio of dressed human body surface area to the naked human body surface area, t_{cl} is outer surface temperature of clothing (°C), t_r is mean radiant temperature (°C), h_c is convective heat transfer coefficient of clothing outer surface (W/(m^2 K)).

PMV can predict the thermal sensation of a group of people in a given thermal condition. But due to large individual difference in thermal sensation between people, the statistical PPD index was proposed to predict the proportion of unsatisfied people in a given environment. In general, there is a quantitative relationship between PMV and PPD, as shown in Eq. 1.2.

$$PPD = 100 - 95^{\{-(0.03353PMV^4 + 0.2179PMV^2)\}} \quad (1.2)$$

Figure 1.11 shows the comfort zone of indoor thermal environment based on PMV calculation in ASHRAE Standard 55. In summer, when the clothing insulation is 0.5 clo and the relative humidity is 50%, the comfortable temperature range is 24–27 °C. In winter, when the clothing insulation is 1.0 clo and the relative humidity is 50%, the indoor comfortable temperature ranges from 20.5 to 24.5 °C.

Based on PMV-PPD calculation, many comfort evaluation standards classify indoor thermal conditions into different grades. The higher the grade, the narrower the range of PMV is allowed to deviate from thermal neutrality, leading to a stricter indoor temperature limit. As shown in Table 1.2, the grade I in ISO7730 requires to control the indoor temperature constantly, only 0.2 PMV fluctuation is allowed.

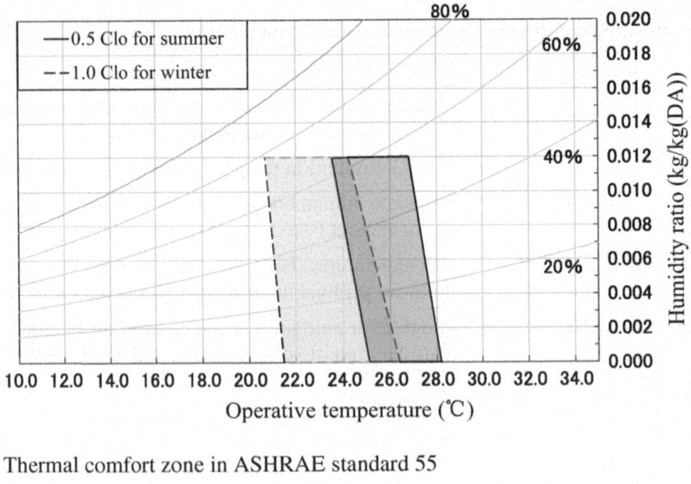

Fig. 1.11 Thermal comfort zone in ASHRAE standard 55

Table 1.2 Thermal environment classifications

Standard	Class	PMV range	PPD range (%)
ISO 7730 [12]	I	$-0.2 < \text{PMV} < +0.2$	<6
	II	$-0.5 < \text{PMV} < +0.5$	<10
	III	$-0.7 < \text{PMV} < +0.7$	<15
EN 15251 [14]	I	$-0.2 < \text{PMV} < +0.2$	<6
	II	$-0.5 < \text{PMV} < +0.5$	<10
	III	$-0.7 < \text{PMV} < +0.7$	<15
	IV	$\text{PMV} < -0.7$ or $\text{PMV} > +0.7$	>15
ASHRAE 55 [13]	–	$-0.5 < \text{PMV} < +0.5$	<10
GB/T 50785 [15]	I	$-0.5 < \text{PMV} < +0.5$	<10
	II	$-1 < \text{PMV} < -0.5$ or $+0.5 < \text{PMV} < 1$	10–25
	III	$\text{PMV} < -1$ or $\text{PMV} > 1$	>25

Grade II allows a slightly larger PMV fluctuation range of 0.5 scale unit. But both grade I and II require indoor air movement speed to be limited under the human sensitive threshold (<0.2 m/s).

1.2.2.2 Rethinking About the PMV

In early thermal comfort studies, the PMV-PPD model provided a quantitative way to predict occupants' thermal sensation and thermal satisfaction. It played an important role in guiding and promoting HVAC technologies. However, the model is based on steady-state uniformly conditioned experiments, which leads to a large discrepancy between the model prediction and the actual feelings. The following aspects have been investigated regarding the limitations of the PMV model.

(1) The accuracy of the PMV model needs to be enhanced

The accuracy of the PMV model has been widely investigated in thermal comfort field. Since the appearance of the model, researchers started to verify its accuracy through climate chamber experiments and field investigations. Many on-site surveys showed that building occupants show great adaptability and tolerance to thermal conditions deviated from neutrality. And that people lived in non-air-conditioned buildings can accept wider indoor temperature range than those lived in air-conditioned buildings [58–60]. Table 1.3 lists part of these studies. It indicates that factors such as regional location, climate condition, and building operation mode can influence building occupants' acceptable temperature range.

For better predicting thermal comfort in non-air conditioning buildings, Fanger et al. [75] extended the PMV model by introducing thermal expectation index. They collected multiple field study data and then compared the PMV prediction with the

Table 1.3 Field studies on thermal comfort

Researcher	Location	Season	Building operation mode	Sample size	80% acceptable temperature range (°C)
Humphreys [16]	European	Whole year	NV	27284	20.1–25.3
Paliaga [61]	U.S.	Summer	NV	1045	19.7–26.3
		Winter	NV	1030	16.8–27.4
de Dear [62]	Singapore	Whole year	NV	2807	20.4–25.1
Candido [63]	Brazil	Whole year	NV	2075	25.3—31.4
Mustapa [64]	Japan	Summer	NV	401	24.6–28.7
Indraganti [19]	India	Summer	NV	3960	26–32.4
Wang [65]	Harbin, China	Winter	Centralized heating	120	18.0–25.5
Zhang [66, 67]	Harbin China	Summer	NV	1285	21.5–31.0
Xia [68]	Beijing China	Summer	NV	88	22.2–30.0
Yang [69]	Xi'an China	Summer	NV	–	26.0–30.7
Li [70]	Nanyang China	Summer	NV	1320	24.2–29.8
Liu [71]	Chongqing China	Whole year	NV	3621	15.4–27.1
Jiang [72]	Changsha China	Summer	NV	523	18.2–31.8
Yang [73]	Wuhan China	Summer	NV	129	25.0–31.6
			AC	100	25.1–30.3
Chen [74]	Guangzhou China	Summer	NV	501	25.2–30.9

actual survey data. It showed that there is a 'scissors difference' between the PMV and the actual surveys, especially in warm conditions (shown in Fig. 1.12).

(2) Indoor thermal environment grading

Table 1.2 summarizes the indoor thermal environment grading method in different standards. Basically, the higher the class, the narrower PMV range should be controlled. But can these higher classes obtain better comfort evaluation in real buildings? Arens et al. [76] analyzed the subjective feedbacks in ASHRAE RP884 database [77], and found that there was no significant satisfaction rate difference between Class I and Class II/III environment, all being around 80%.

The basic logic behind the grades is the assumption that the more stable and uniform the indoor environment is, the more thermally comfortable occupants should be. However, this assumption is not necessarily true. Many researchers found that thermal neutrality is equivalent to thermal comfort. Comfort feeling often comes

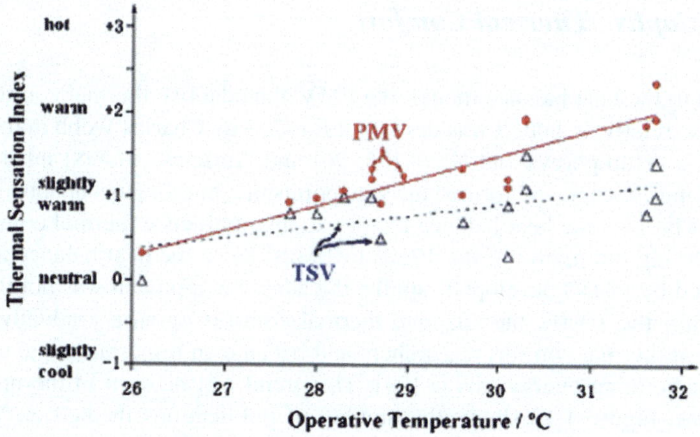

Fig. 1.12 Comparison of PMV and actual observations [75]

from the elimination of discomfort. Even when our body is in the unneutral thermal state, we can still have a comfortable feeling [78–80]. Based on this phenomenon, de Dear et al. [81, 82] revisited the 'alliesthesia' concept with the holding that introducing proper cold/hot stimulus can broaden the thermal comfort range. The Tsinghua research team also have been advocating the 'dynamic thermal comfort' to encourage natural ventilation strategies [83].

(3) Potential health issues of 'steady-state' thermal environment

With the popularization of HVAC, relevant health issues have attracted researchers' attention. Many studies show that steady-state neutral thermal conditions consumes a large amount of energy but not necessarily the most comfortable and healthy [84]. The Brown Adipose Tissue (BAT) in human body investigated by van Marken Lichtenbelt [85, 86] is one example. Their studies found that proper deviation from thermal neutrality is beneficial to human health [87]. Johnson [88] and Hippel [89] analyzed the relationship between thermal environment and the obesity epidemiology. It was found that long-term living in thermal neutrality is an important factor causing the obesity epidemic in Europe and America. Jay et al. [90] compared the effects of the frequent thermal stimulus on residents' thermo-physiological regulation capability. They found that excessive use of air conditioning and lack of thermal stimulus would reduce occupants' capability to adapt to a warm environment. Similar phenomea were reported by Chinese researchers [91] as well.

Human beings have a long history of evolution. We learned to maintain thermal comfort through building envelop insulation, natural ventilation, clothing adjustment, and other ways, which reflect the adaptability to the natural environment. The air conditioning technology occurred in less than 100 years, but it has dramatically shaped today's building environment. Advances in HVAC technologies allow us to live in thermally neutral building environments with no thermal stress, but it may also cause discomfort symptoms such as 'sick building syndrome' and 'air conditioning disease' [92–94].

1.2.3 Adaptive Thermal Comfort

Compared to the heat balance model like PMV, the adaptive thermal comfort came into popular relatively late. It was firstly put forward by Charles Webb in the 1960s. In the 1970s, Humphreys and Nicol [95, 96] and Auriems [97, 98] inherited and developed the concept, and formed the viewpoints that building occupants' behavior adjustment besides the heat balance factor can also influence thermal comfort perception. During the period of the 1960s to 1980s, since the heat balance approach (represented by PMV) developed rapidly, the adaptive approach was not the mainstream. Until the 1990s, the adaptive thermal comfort opinion gradually entered the vision of thermal comfort researchers and became an important topic under the impetus of building energy-saving [99]. This trend can be seen in the number of publications. Figure 1.13 shows the number of publications themed on 'adaptive thermal comfort' in China and Other countries. During the 27 years from 1970 to 1997, only 6 articles were included in the core collection of Web of Science, while 379 articles were published from 1998 to 2015 with a year-by-year increasing trend. Chinese researchers entered this field relatively late but increased rapidly. Especially in recent years, the number of publications from China accounted for roughly 15% of the total papers.

1.2.3.1 Theoretical Frameworks

Different from the heat balance approach, which emphasizes the influence of the thermal environment on human body heat transfer, the adaptive thermal comfort pays more attention to the role played by building occupants. It is believed that building occupants' thermal comfort can be influenced by climate condition, psychological expectation, cultural background, and other factors. All these factors together formed several adaptive layers, such as behavior adjustment, physiological acclimatization

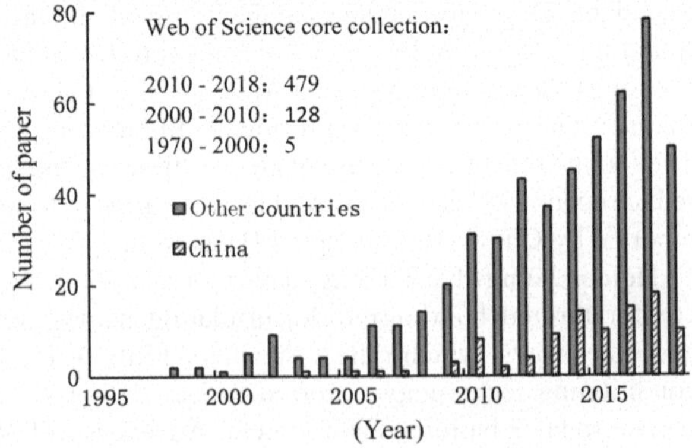

Fig. 1.13 Number of papers themed on 'thermal comfort' and 'adaptive thermal comfort' on Web of Science (accessed on 2015-12-20)

and psychological adaptation [100–103]. It is believed that subjects in climate chamber experiments underpinning the PMV model can only accept the experimental conditions passively so that the model can only reflect the one-way influence of thermal conditions on subjective feelings but cannot reflect the mutual feedbacks between actual building environment and occupants' perception. Unlike the heat balance comfort models, the adaptive approach emphasizes the interactions between building occupants and the building environment (shown in Fig. 1.14). It holds that when people feel uncomfortable or dissatisfied with their surrounding thermal environment, it's not the end of the loop, but the beginning of adaptation.

A more explicit adaptive thermal comfort framework was proposed by Chinese researchers in 2010 [104] (as shown in Fig. 1.15). The new framework assumes that human body heat balance status is a starting point for the adaptive cycles, and

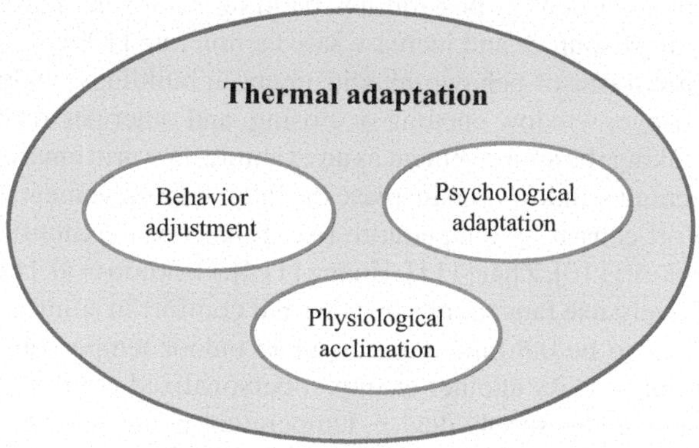

Fig. 1.14 Adaptive thermal comfort theory proposed by de Dear and Brager [99, 103]

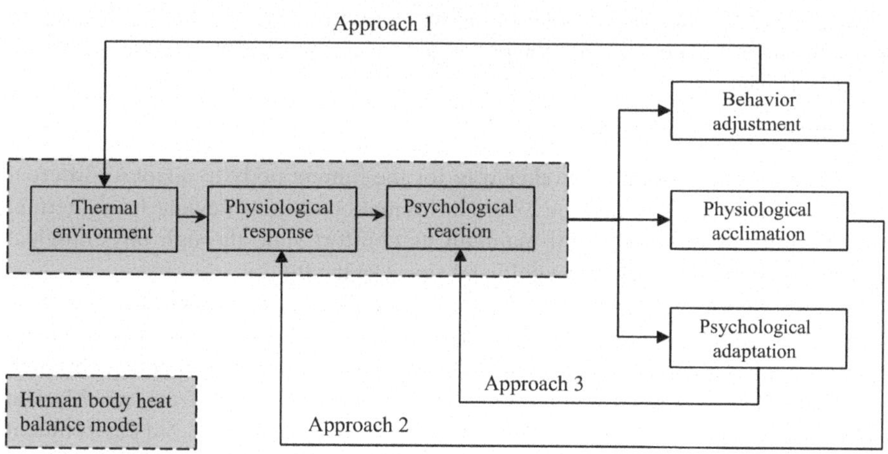

Fig. 1.15 A framework for adaptive thermal comfort theory [104]

explicitly gives the processes of three cycles (e.g. behavior adjustment, physiological acclimation, and psychological adaptation). It integrates the adaptive approaches with the heat balance model and points out a way to establish a more comprehensive adaptive thermal comfort model.

(1) Behavioral adjustment

Behavior adjustment directly acts on the thermal environment. It has the most obvious effect on thermal comfort and can be observed easily. In real buildings, occupants' thermal demand may vary greatly because of factors such as gender [105], age [106], dress custom [107], activity intensity [108]. Behavior adjustment can easily compensate such individual differences by putting on more clothes, turning on a local fan, or opening the windows. When the thermal environment deviates from neutrality, various measures can be taken to maintain human body in a thermal neutral state. The more environment control possibilities building occupants have, the easier to neutralize discomfort sources and increase satisfaction rate [109].

There are many forms of behavioral adjustment in buildings, including clothing adjustment, fan usage, window opening or closing, and other personal comfort system (PCS), etc. Taking the air movement as an example, in warm environment, proper use of fans or opening windows can increase the indoor air movement, thus offsetting thermal discomfort caused by temperature rise. It has been demonstrated by many researchers. Melikov [110], Zhai [111], Huang [112], Candido et al. [113] that people are willing to actively use fans to improve thermal comfort in summer. If the indoor airspeed is allowed to be 0.8 m/s, the acceptable indoor temperature can be lifted to 30 °C or even higher. As another example, personalized comfort system (PCS), including personalized air supply device, temperature-controlled seat, heated desktop, etc. can provide building occupants more flexible behavior adjustment, which can not only reduce HVAC energy consumption [114, 115], but also maximize thermal comfort [116]. Arens and Zhang et al. [117] found that by using PCS, users can achieve over 85% satisfaction rate in the temperature range of 16–30 °C. Even in thermally neutral zone, PCS has the potential to increase satisfaction rate from 70% to nearly 100%.

(2) Physiological acclimation

Physiological acclimation is another way for the human body to adapt to different thermal environments. When the external thermal conditions deviate from thermal neutrality, the human body will maintain its comfort state through physiological regulations, such as shiver in the cold and sweat in the hot.

In recent years, many researchers reported the role of physiological acclimation played in thermal adaptation. One example is brown adipose tissue (BAT) related research. Earlier studies believed that BAT only exists in infants and young children, adults only have a limited amount of BAT. But recent findings from van Marken Lichtenbelt's team [118] showed that exposing subjects to a mild cold environment for (14–15 °C) 6 h a day can improve adults' BAT activity after only 10 days, thus improving the metabolic rate regulation and thermal comfort evaluation. They also

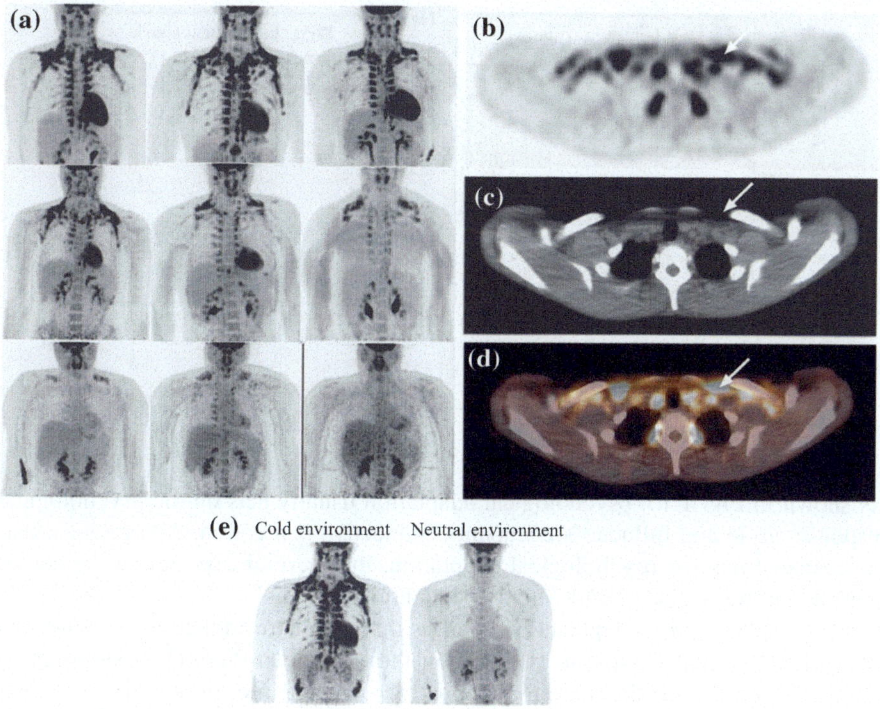

Fig. 1.16 Brown tissue activity in different thermal conditions [85]

found that cold exposure is helpful to improve insulin sensitivity of type 2 diabetes patients [119], and human body metabolic rate will increase significantly in both cold and hot exposures [120] (Fig. 1.16).

Chinese researchers also made some findings in this field. In a comparative experiment, Yu [121] found that people lived in the air-conditioning environment for a long period not only had less tolerance to hot environment, but also had lower heat stress protein HSP70 level in their blood (shown in Fig. 1.17). Cao [67] found that human body physiological acclimation can be divided into three categories based on the length of adaptation time. The first one is caused by sudden changes in the external microenvironments, such as entering a heated room from a cold outdoor in winter or leaving an air-conditioned room to a hot outdoor in summer. The second level changes alternately with seasons. For example, people often have different requirements for indoor temperature in different seasons. In summer, people can accept higher room temperature, while in winter they may feel comfortable at lower room temperature. The third level is caused by long-term acclimatization, such as moving from one climate zone to another, which may need several years to adapt.

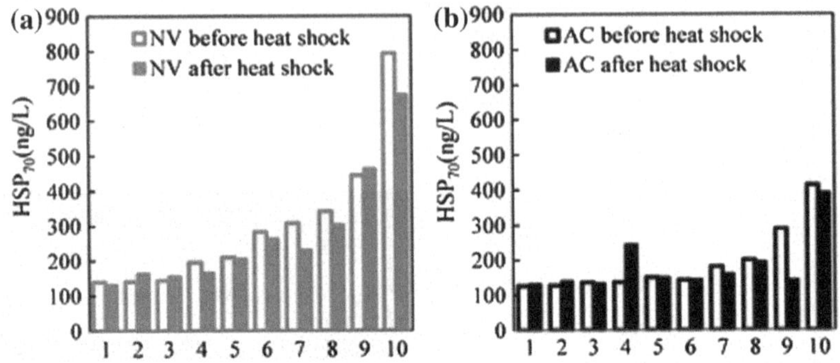

Fig. 1.17 HSP70 levels pre and post heat shock [121]

(3) Psychological adaptation

As shown in Fig. 1.15, psychological adaptation usually acts on the psychological response circle and influences occupants' subjective perception. Among the existing explanations for psychological adaptation, the thermal expectation, perceived control, and forgiveness factor should be mentioned.

Thermal expectation explanation believes that people form their expectations on the current thermal environment based on their previous thermal experiences. If people live in a poor thermal environment for a long time, they will lower their requirements, thus lowering their corresponding expectations. Contrarily, if people live in a comfortable thermal environment for a long time, they will gradually raise their requirements. Correspondingly, their expectations on the thermal environment will be higher. The thermal expectation explanation was firstly proposed by Fanger and Toftum [122] and was applied to explain why PMV over estimated actual TSV in non-air-conditioned buildings. An expectation factor *e* depending on climate condition and building operation mode was introduced to modify the PMV (as shown in Eq. 1.3). Its values generally ranged from 0.5 to 1. The modified PMV_e agrees with actual TSV much better and has been utilized in other studies [123, 124].

$$PMV_e = e \times PMV \quad (1.3)$$

Perceived control mainly emphasizes building occupants' control capability on the thermal environment. It believes that whether there is control to the thermal environment will affect occupants' thermal comfort. For example, if occupants have higher control over discomfort sources, their negative emotions would be lower and their acceptable temperature range will be wider [26]. In a psychological experiment, Zhou [125] found that when subjects lost control ability to the surrounding thermal environment, their TSV deteriorated by 0.4–0.5 scale unit.

Forgiveness factor measures people's tolerance to an uncomfortable environment. When the forgiveness factor is greater than 1, the user is more tolerant. When the forgiveness factor is less than 1, it means the user is picky. Through analyzing the survey

Table 1.4 Forgiveness factor in buildings with different operation mode [136]

Index	Naturally ventilated buildings	Mixed-mode buildings	Air-conditioned buildings
Forgiveness factor	1.15	1.10	1.05
Building samples	41	21	64

data of 177 buildings, Bordass [126, 127] found that building users have different forgiveness factors for buildings with different operation modes. As shown in Table 1.4, occupants' forgiveness factors for natural ventilation and mixed-mode buildings are 1.15 and 1.10 respectively, which are larger than the 1.05 of air-conditioned buildings. It suggests that occupants are more willing to tolerate the discomfort in naturally ventilated buildings, but easier to complain about the discomfort in air-conditioned buildings.

1.2.3.2 Data Collection

As an important supplement to climate chamber experiment, field investigation is another widely used method in thermal comfort field. Compared with the climate chamber experiment method, field investigation in real buildings can better capture building occupants' actual feelings. Therefore, establishing a detailed building thermal comfort database can be a very useful tool for thermal comfort research. In recent years, thermal comfort field studies have been carried out in different climates and different buildings types. These studies accumulated a large number of samples in real buildings, with more than 100,000 samples. Table 1.5 lists part of these studies.

To date, there are three available databases that have a wide impact. One is the SCATs database [128] established in 1997. It contains 4655 data sets from 26 buildings in 5 European countries. It provides the foundation for the adaptive thermal comfort model in European EN15251 standard. The other one is the ASHRAE RP-884 database [77] established in 1998. It covers more than 21,000 samples from 160 buildings on four continents and provides a basis for the adaptive model in ASHRAE standard 55. In 2017, a new global thermal comfort database, including 6 continents, 23 countries, and nearly 81846 samples, has been launched (as shown in Fig. 1.6). It provides a basis for further thermal comfort data analysis.

In China, researchers have carried out field investigations on thermal comfort in different climate zones and different building types. Cao [59] studied the seasonal and regional characteristics of Chinese people's thermal adaptation and proposed new thermal adaptation model. Yu [94, 143] compared occupants' thermal adaptation and thermal health with different thermal experiences. She found that staying in a stable thermal neutral environment may have negative effect on building occupants' health. Yang [40, 144] conducted large-scale thermal comfort studies in typical cities with different climate conditions. Li and Yao [145–147] put forward the adaptive PMV

Table 1.5 Part of thermal comfort field studies

Researcher	Location	Building types	Season	Building operation	Sample size
Humphreys [129]	European	Office	Whole year	NV + AC	4655
Toftum [130]	Denmark	Teaching building	Winter	AC	170
Singh [131]	Belgium	Residential	Winter	NV	85
Paliaga [61]	The U.S.	Office	Summer	NV	1045
			Winter	NV	1030
Kim [132]	The U.S.	Office	Winter	AC	85
Langevin [133]	The U.S.	Office	Summer	AC	570
de Dear [134]	Australia	Office	Whole year	NV + AC	1335
Candido [63, 135]	Brazil	Teaching building	Spring	MM	5036
Romero [136]	Mexico	Resident	Winter	NV	1424
Mustapa [64]	Japan	Teaching building	Summer	NV	401
Rijal [137]	Japan	Residential	Summer	NV	8280
				AC	4860
Madhavi [138]	India	Office	Whole year	NV + AC	6048
Myla [139]	Philippian	Office	Whole year	AC	277
Heidari [140]	Iran	Office	Whole year	NV	512
Oseland [141]	South Africa	Office	Whole year	NV	20997
Bouden [142]	Tunisia	Office	Summer	NV	1651

model (aPMV = PMV/1 + λPMV, where λ is the adaptive coefficient) by combining the 'adaptive thermal comfort' with the PMV model. Wang [148, 149] investigated thermal adaptation in cold areas and developed adaptive models for that area. Zhang [150–153] suggested that the 'black-boxed' adaptive model should be 'white-boxed'. Figure 1.18 shows a glimpse of such field studies conducted in China.

1.2.3.3 Adaptive Comfort Models and Standards

With the advances in adaptive thermal comfort theory and data collection, several models have been established to guide the design and operation for relevant buildings (especially for naturally ventilated buildings). A classic example is the model (as shown in Eq. 1.4) developed by de Dear and Brager [99], which has been adopted by ASHRAE standard 55. The other one was developed by Humphreys and Nicol [128] which is based on SCATs database and has been incorporated in EN15251 standard. These two models are similar, except that their regression coefficients and outdoor temperature formats are slightly different.

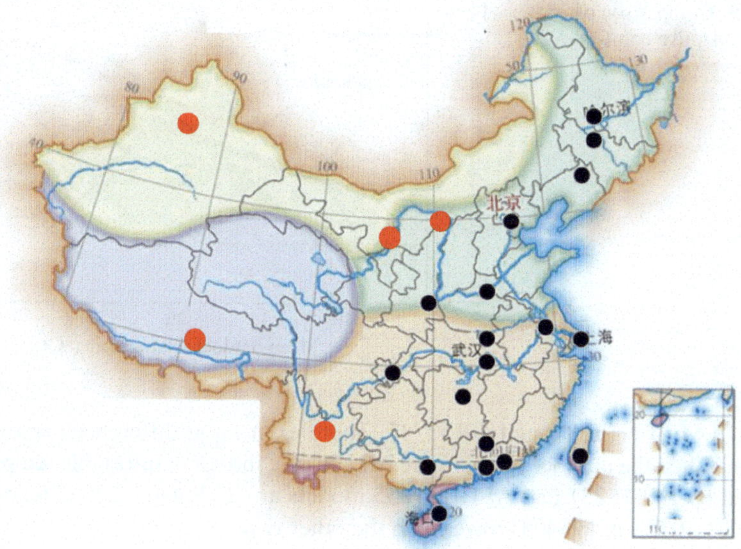

Fig. 1.18 Locations of thermal comfort field studies in China. *Note* The figure was modified based on the work of Zhang and Zhao [104]

$$t_{comf} = 0.31t_{out,m} + 17.8 \tag{1.4}$$

where t_{comf} is indoor comfortable temperature (°C), $t_{out,m}$ is monthly mean outdoor temperature (°C).

In Chinese GB50785 standard, the adaptive PMV (aPMV) [154] developed by Li and Yao was adopted. It was shown in Eq. 1.5. The λ in the formula is called the adaptive coefficient, which reflects the degree of different populations' thermal adaptation. Table 1.6 shows the detailed values.

$$aPMV = PMV/(1 + \lambda PMV) \tag{1.5}$$

It should be noted that most of the existing adaptive models are based on 'black-boxed' linear regression. Figure 1.19 compares the existing adaptive comfort models

Table 1.6 The adaptive coefficient in different building types [15]

Climate region		Residential, commercial, hotel, and office	Teaching buildings
Sever cold, cold	$PMV \geq 0$	0.24	0.21
	$PMV < 0$	−0.50	−0.29
Hot summer cold winter, hot summer warm winter, mild	$PMV \geq 0$	0.21	0.17
	$PMV < 0$	−0.49	−0.28

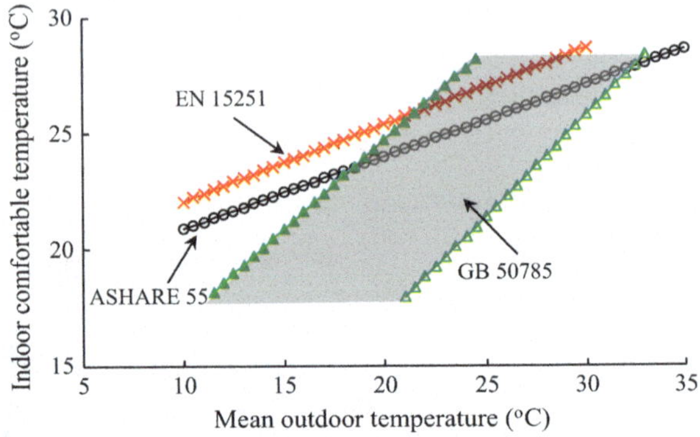

Fig. 1.19 Comparison of existing adaptive comfort models

in ASHRAE, EN15251, and GB50785. There exist large differences among these models. When the outdoor temperature is 15 °C, the indoor comfortable temperatures required by EN 15251 and ASHRAE 55 models are 23.8 and 22.5 °C respectively, while GB 50785 is 19, 4 °C lower than the other two.

1.2.4 Summary and Comment

To sum up, the heat balance comfort models represented by PMV has played an important role in guiding and promoting HVAC technology. But many studies showed that these models deviated greatly from the actual comfort perception in real buildings. The adaptive thermal comfort approach emphasizes the interactions between building occupants and environment control. It also emphasizes the correlation between indoor thermal comfort and outdoor climate conditions. To some extent, it promotes climatic building design and environmental control strategies. But there remain some issues to be explored.

The current adaptive models didn't capture the adapting process. Most existing adaptive thermal comfort studies focused on long-term adaptation phenomenon but rarely mention the adapting process, especially the time scale of these processes. As a result, the current adaptive models heavily rely on field survey data while being unable to predict future situations. In addition, the flaws of existing thermal comfort database make the adaptive models not representative enough. For example, the two main data platforms (SCATs database and RP-884 database) were established in the 1990s. They are limited by sample sizes, data uniformity, sample representativeness, and time validity. Given the fact that the database method cannot enumerate all future scenarios, specifying its essential characteristics can help achieve a better understanding of thermal adaptation in buildings.

The mechanism behind thermal adaptation has been absent. Existing theories explained thermal adaptation from three perspectives of behavioral adjustment, physiological acclimatization, and psychological adaptation, but they remain in assumptions. Few of these approaches have been verified, thus leaving behind many disputed questions such as 'Do physiological acclimatization and psychological adaptation really exist?' 'How effective are they?' 'Is it possible to quantify their effects?' If a 'white-box' adaptive model with clear mechanism is going to be built, the effects of these assumed adaptive layers should be re-visited.

Lacking models with clear mechanism. As the adapting process and the underlying causes of thermal adaptation remain unclear, it is difficult to quantify the effects of each adaptive approach. Consequently, most existing adaptive comfort models used 'black-box' statistical regression without integrating the adaptive approaches with the heat balance models (such as PMV), making these two modeling approaches incompatible with each other. If the essential characteristics of thermal adaptation and its underlying mechanisms can be captured, it is promising to develop more comprehensive models with both the computational advantages of heat balance model and empirical advantages of the statistical model.

1.3 Research Design

1.3.1 Content and Technical Route

The research of this thesis aims to reveal the essential characteristic and underlying mechanism of building occupants' thermal adaptation. To achieve this goal, a comprehensive method with the online surveys, field investigations, climate chamber experiments, and model analysis is applied. The following studies were conducted throughout this research.

(1) First, we reviewed the cutting edge of adaptive thermal comfort research, especially from the perspectives of human body heat balance comfort model and adaptive thermal comfort model (Chap. 1).
(2) Second, we conducted an online survey in China, where wintertime indoor thermal environments in northern cities (with district heating) is much warmer than in southern region (without district heating), to investigate the mutually dependent relationship between indoor climate experience and occupants' comfort expectation (Chap. 2).
(3) Third, we carried out a comparative thermal comfort field investigation to depict the dynamic processes and time-scale of building occupants' thermal adaptation to different indoor climates (Chap. 3).
(4) Forth, we conducted a series of climate chamber experiment to understand how physiological acclimation affects building occupants' thermal adaptation and quantitatively described the effects (Chap. 4).
(5) Fifth, we investigated the psychological effects of personal control through field investigations and comparative laboratory experiments. The psychological effect of perceived control was described in a quantitative way (Chap. 5).
(6) Lastly, we developed the adaptive heat balance model (AHBM) to connect the adaptive comfort framework and the heat balance comfort model (Chap. 6).

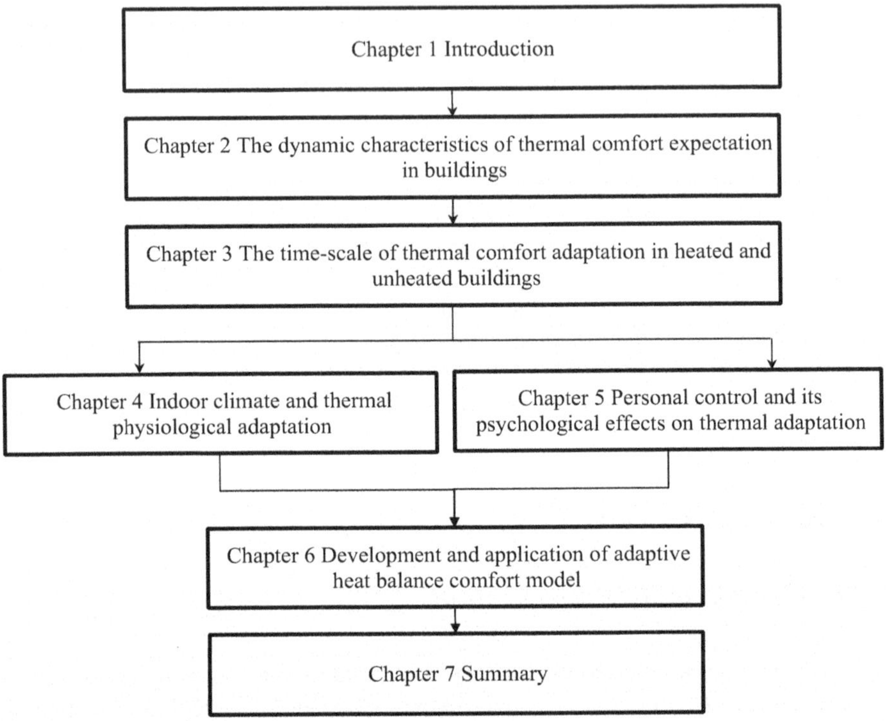

Fig. 1.20 Thesis framework

1.3.2 Thesis Framework

The framework of this thesis is shown in Fig. 1.20.

References

1. Tsinghua University Building Energy Conservation Research Center (2015) China building energy conservation annual development research report, Beijing
2. National Bureau of Statistics of China (2018) China statistical yearbooks. China Statistical Press, Beijing, 1990–2014. Available at: http://data.stats.gov.cn. Accessed 26 Apr 2018
3. Yang L (2010) Architectural climatology. China Building Industry Press, Beijing
4. Yang L (2003) Research on building climate analysis and design strategy. Ph.D. thesis. Xi'an University of Architecture and Technology, Xi'an
5. Lin Y (2014) Climate adaptation of China's climate and human thermal comfort. J Xi'an Univ Archit Technol 46(2):251–255
6. Davis L, Gertler P (2015) Contribution of air conditioning adoption to future energy use under global warming. Proc Natl Acad Sci USA 112(19):5962–5967
7. Energy Information Administration (EIA) (2018) Residential energy consumption survey. Available at: http://www.eia.doe.gov. Accessed 26 Apr 2018

8. Statistics Japan, National Survey of Family Income and Expenditure (2018) Available at: http://www.stat.go.jp/english/index.html. Accessed 26 Apr 2018
9. Akpinar-Ferrand E, Singh A (2010) Modeling increased demand of energy for air conditioners and consequent CO_2 emissions to minimize health risks due to climate change in India. Environ. Sci. Pol. 13(8):702–712
10. Wikipedia (2016) Air conditioning: development of mechanical cooling (2016-01-26). https://en.wikipedia.org/wiki/Air_conditioning#cite_note-13
11. Fanger PO (1970) Thermal comfort. Analysis and application in environment engineering. Danish Technology Press, Copenhagen
12. Technical Committee ISO/TC 159 and Technical Committee CEN/TC 122. ISO 7730 (2005) Ergonomics of the thermal environment—analytical determination and interpretation of thermal comfort using calculation of the PMV and PPD indices and local thermal comfort criteria. European Committee for Standardization, UK
13. ASHRAE (2013) Thermal environmental conditions for human occupancy. ASHRAE Standard 55, American Society of Heating, Refrigerating and Air-Conditioning Engineers, Atlanta, Georgia
14. CEN (2007) Indoor environmental input parameters for design and assessment of energy performance of buildings: addressing indoor air quality, thermal environment, lighting and acoustics. EN 15251, European Committee for Standardization, Brussels
15. GB/T (2012) Evaluation standard for indoor thermal environment in civil buildings, GB/T 50785. Ministry of Housing and Urban-Rural Development of the People's Republic of China, Beijing (in Chinese)
16. Humphreys MA (1976) Field studies of thermal comfort compared and applied. J Inst Heat Vent Eng 44(1):5–27
17. Emmerich S, Polidoro B, Axley J (2011) Impact of adaptive thermal comfort on climatic suitability of natural ventilation of office buildings. Energy Build 43(9):2101–2107
18. Barlow S, Fiala D (2007) Occupant comfort in UK offices: how adaptive comfort theories might influence future low energy office refurbishment strategies. Energy Build 39:837–846
19. Indraganti M (2010) Using the adaptive model of thermal comfort for obtaining indoor neutral temperature: findings from a field study in Hyderabad, India. Build Environ 45:519–536
20. Loonen R, Treka M, Costola D, Hensen J (2013) Climate adaptive building shells: state-of-the-art and future challenges. Renew Sustain Energy Rev 25:483–493
21. de Dear RJ, Brager GS (2002) Thermal comfort in naturally ventilated buildings: revision to ASHRAE Standard 55. Energy Build 34:549–561
22. Luo M, Cao B, Damiens J, Lin B, Ouyang Q et al (2015) Evaluating thermal comfort in mixed-mode buildings: a field study in a subtropical climate. Build Environ 88:46–54
23. Deuble MP, de Dear RJ (2012) Mixed-mode buildings: a double standard in occupants' comfort expectation. Build Environ 54:53–60
24. Halawa E, van Hoof J (2012) The adaptive approach to thermal comfort: a critical overview. Energy Build 51:101–110
25. de Dear R, Foldvary V, Zhang H, Arens E, Luo M et al (2016) Comfort is in the mind of the beholder: a review of progress in adaptive thermal comfort research over the past two decades. In: Proceedings of the 5th International Conference on Human-Environment System, Nagoya, Japan
26. Nikolopoulou M, Steemers K (2003) Thermal comfort and psychological adaptation as a guide of designing urban spaces. Energy Build 35(1):95–101
27. Chen L, Ng E (2012) Outdoor thermal comfort and outdoor activities: a review of research in the past decade. Cities 29(2):118–125
28. Akimoto T, Tanabe S, Yanai T, Sasaki M (2010) Thermal comfort and productivity—evaluation of workplace environment in a task conditioned office. Build Environ 45(1):45–50
29. Cui W, Cao G, Ouyang Q, Zhu Y (2013) Influence of dynamic environment with difference airflows on human performance. Build Environ 62(62):124–132
30. Lan L, Lian Z, Pan L (2010) The effects of air temperature on office worker's well-being, workload and productivity—evaluated with subjective ratings. Appl Ergon 42(1):29–36

31. Yang L, Yan H, Lam J (2014) Thermal comfort and building energy consumption implication—a review. Appl Energy 115(4):164–173
32. Kwong Q, Adam N, Sahari B (2014) Thermal comfort assessment and potential for energy efficiency enhancement in modern tropical buildings: a review. Energy Build 68(314):547–557
33. Taleghani M, Tenpierik M, Kurvers S et al (2013) A review into thermal comfort in buildings. Energy Rev 26(10):201–215
34. Zomorodian Z, Tahsildoost M (2016) Thermal comfort in educational buildings: a review article. Sustain Energy Rev 56:895–906
35. Khodakarami J, Nasrollahi N (2012) Thermal comfort in hospitals—a literature review. Renew Sustain Energy Rev 16(6):4071–4077
36. Cui W, Wu T, Ouyang Q, Zhu Y (2016) Passenger thermal comfort and behavior: a field investigation in commercial aircraft cabins. Indoor Air
37. Lan L, Lian Z (2016) Ten questions concerning thermal environment and sleep quality. Build Environ 99:252–259
38. de Dear R, Akimoto T, Arens E, Brager G et al (2013) Progress in thermal comfort research over the last twenty years. Indoor Air 23(6):442–461
39. Rupp R, Vasquez N, Lamberts R (2015) A review of human thermal comfort in built environment. Energy Build 105:178–205
40. Huang J, Zhang H (2011) Human and thermal environment. Science Press, Beijing
41. Mclntyre DA (1978) Seven-point scales of warmth. Build Sci Eng 45:215–226
42. Zhang Y, Wang J, Chen H, Zhang J (2010) Thermal comfort in naturally ventilated buildings in hot-humide area. Build Environ 45(11):2562–2570
43. Foldvary V, Cheung T, Zhang H et al (2018) Development of the ASHRAE global thermal comfort database II. Build Environ 142:502–512
44. Gagge AP, Stolwijk JA, Hardy JD (1967) Comfort and thermal sensations and associated physiological responses at various ambient temperatures. Environ Res 1:1–20
45. Berglund LG (1979) Thermal acceptability. ASHRAE Trans 85(2):825–834
46. Fanger PO (1967) Calculation of thermal comfort: introduction of a basic comfort equation. ASHRAE Trans 73III.4.1–III.4.20
47. Araújo VMD, Araújo EHS (1999) The applicability of ISO 7730 for the assessment of the thermal conditions of users of the buildings in Natal-Brazil. Proc Indoor Air 2:148–153
48. Yoon DW, Sohn JY, Cho KH (1999) The comparison on the thermal comfort sensation between the results of questionnaire survey and the calculation of the PMV values. Proc Indoor Air 2:137–141
49. Mayer E (1997) A new correlation between predicted mean votes (PMV) and predicted percentages of dissatisfied (PPD). Proc Healthy Build 2:189–194
50. Zhu Y et al (2016) Built Environment, 4th edn. Beijing, China Building Industry Press
51. Madsen T, Olesen B, Kristensen N (1984) Comparison between operative and equivalent temperature under typical indoor conditions. ASHRAE Trans 90(1):1077–1090
52. Gagge A, Fobelets A, Berglund L (1987) A standard predictive index of human response to the thermal environment. ASHRAE Trans 92:709–731
53. Ye G, Yang C, Chen Y, Li Y (2003) A new approach for measuring predicted mean vote (PMV) and standard effective temperature (SET). Build Environ 38(1):33–44
54. Huang L, Arens E, Zhang H, Zhu Y (2014) Applicability of whole-body heat balance models for evaluating thermal sensation under non-uniform air movement in warm environments. Build Environ 75:108–113
55. Jendritzky G, de Dear R, Havenith G (2012) UTCI—why another thermal index? Int J Biometeorol 56:421–428
56. Budd G (2008) Wet-bulb globe temperature (WBGT)-its history and its limitations. J Sci Med Sport 11(1):20–32
57. Schiavon S, Hoyt T, Piccioli A (2013) Web application for thermal comfort visualization and calculation according to ASHRAE Standard 55. Build Simul 7(4):321–334
58. Zhou X (2008) Study on influencing factors and evaluation indexes for human thermal sensation in warm conditions. Ph.D. thesis. Tsinghua University, Beijing

59. Cao B (2012) Research on the impacts of climate and built environment on human thermal adaptation. Ph.D. thesis. Tsinghua University, Beijing
60. Yan H (2013) Study on adaptive thermal comfort on the basis of regions and climates of China. Ph.D. thesis. Xi'an University of Architecture Technology, Xi'an
61. Brager G, Paliaga G, de Dear R (2004) Operable windows, personal control and occupant comfort. ASHRAE Trans 110(2):17–35
62. de Dear R, Leow K, Foo S (1991) Thermal comfort in the humid tropics: field experiments in air conditioned and naturally ventilated buildings in Singapore. Int J Biometeorol 34(4):259–265
63. Candido C (2010) Indoor air movement acceptability and thermal comfort in hot-humid climates. Ph.D. thesis. Macquarie University, Sydney
64. Mustapa M, Zaki S, Rijal H, Hagishima A, Ali M (2016) Thermal comfort and occupant adaptive behavior in Japanese university buildings with free running and cooling mode offices during summer. Build Environ 105(15):332–342
65. Wang Z, Fang X, Lian L (2002) Field experiments on occupant thermal comfort in Harbin. J Harbin Inst Technol 34(4):500–504
66. Zhang L (2010) Study on adaptation and thermal comfort of residents in Harbin. Master thesis. Harbin Institute of Technology, Harbin
67. Wang Z, Zhang L, Zhao J et al (2010) Thermal comfort for naturally ventilated residential buildings in Harbin. Energy Build 42(12):2406–2415
68. Xia Y, Zhao R, Jiang Y (1999) Thermal comfort in naturally ventilated houses in Beijing. HV&AC 29(2):1–5
69. Yang Q (2010) Adaptive thermal comfort model for hot summer and cold winter zone. Master thesis. Xi'an University of Architecture Technology, Xi'an
70. Li J, Yang L, Liu J (2008) Adaptive thermal comfort model for hot summer and cold winter zone. HV&AC 38(7):20–24
71. Liu J (2007) Research on indoor thermal environment and human thermal comfort in natural ventilation buildings in hot summer and cold winter areas. Master thesis. Chongqing University, Chongqing
72. Ye J, Yang C, Li W et al (2006) Gender and thermal comfort in non-air-conditioned environment. HV&AC 36(5):17–21
73. Yang W (2007) Research on thermal comfort adaptation in residential buildings in hot summer and cold winter areas in summer. Master thesis. Hunan University, Changsha
74. Chen H, Zhang Y et al (2010) Thermal comfort in naturally ventilated buildings in hot-humid area of China in summer: an example in Guangzhou. HV&AC 40(2):96–101
75. Fanger PO, Toftum J (2001) Thermal comfort in the future—excellence and expectation. In: Proceedings on moving thermal comfort standards into 21st century, Windsor
76. Arens E, Humphreys M, de Dear R, Zhang H (2010) Are 'class A' temperature requirements realistic or desirable? Build Environ 45:4–10
77. de Dear R (1998) Global database of thermal comfort field experiments. ASHRAE Trans 104:1141–1152
78. Hensel H (1981) Thermoreception and temperature regulation. Academic Press Inc., Burlington
79. Cabanac M (1996) Pleasure and joy, and their role in human life. In: Proceedings of the 7th international conference on indoor air quality and climate, Nagoya
80. Zhao R (2000) Discussion on thermal comfort. HV&AC 30(3):25–26
81. de Dear R (2011) Revisiting an old hypothesis of human thermal perception: alliesthesia. Build Res Inf 39(2):108–117
82. Parkinson T, de Dear R, Candido C (2016) Thermal pleasure in built environments: different thermoregulatory zones. Build Res Inf 44(1):20–33
83. Zhu Y, Ouyang Q, Cao B, Zhou X, Yu J (2016) Dynamic thermal environment and thermal comfort. Indoor Air 26(1):125–147
84. Zhu Y (2015) Thermal comfort: how much is too much. Global Archit 7:35–37
85. van Marken Lichtenbelt W, Vanhommerig J, Smelders N et al (2009) Cold activated brown adipose tissue in healthy men. New Engl J Med 360(15):1500–1508

86. van der Lans A, Hoeks J, Brans B et al (2013) Cold acclimation recruits human brown fat and increases no-shivering thermogenesis. J Clin Investig 123(8):3395–3403
87. van Marken Lichtenbelt W, Schrauwen P (2011) Implications of nonshivering thermogenesis of energy balance regulation in humans. Am J Physiol Regul 301(2):285–296
88. Johnson F, Mavrogiann A, Ucci M et al (2011) Could increased time spent in a thermal comfort zone contribute to population increases in obesity. Obes Rev 12(7):543–551
89. Hippel P, Benson R (2014) Obesity and the natural environment across US countries. Am J Publ Health 104(7):1287–1293
90. Bain A, Jay O (2011) Does summer in a humid continental climate elicit an acclimatization of human thermoregulatory responses? Eur J Appl Physiol 111(6):1197–1205
91. Yu J, Ouyang Q, Zhu Y et al (2012) A comparison of the thermal adaptability of people accustomed to air conditioned environments and naturally ventilated environments. Indoor Air 22:110–118
92. Zhang Y (2013) Indoor air quality control: the challenges and responsibilities of HVAC researchers in the new century (in Chinese). HV&AC 42(12):1–7
93. Tan L, Dai Z, Liu Y (2003) Effects on human thermal feeling and neurobehavioral function in air-conditioning environment. Chin J Publ Health Eng 2(4):193–195
94. Cao B, Shang Q, Dai Z et al (2013) The impact of air-conditioning usage on sick building syndrome during summer in China. Indoor Built Environ 22:490–497
95. Humphreys M (1978) Outdoor temperatures and comfort indoors. Build Res Pract 6(2)
96. Nicol J (1974) An analysis of some observations of thermal comfort in Roorkee, India and Baghdad, Iraq. Ann Hum Biol 1(4):411–426
97. Auliciems A (1969) Effects of weather on indoor thermal comfort. Int J Biometerorol 13:147–162
98. Auliciems A (1982) Psychophysical criteria for global thermal zones of building design. Int J Biometeorol 26(8):69–86
99. de Dear R, Brager G, Readon J, Nicol F (1998) Developing an adaptive model of thermal comfort and preference. ASHRAE Trans 104(1):145–167
100. Mclntyre D (1982) Chamber studies—reduction and absurdum? Energy Build 5:89–96
101. Baker N, Standeven M (1994) Comfort criteria for passively cooled buildings. A PASCOOL task. Renew Energy 5(8):977–984
102. Oseland N (1994) Comparision of the predicted and reported thermal sensation vote in homes during winter and summer. Energy Build 21(1):45–54
103. Brager GS, de Dear RJ (1998) Thermal adaptation in the built environment: a literature review. Energy Build 27(1):83–96
104. Zhang Y, Zhao R (2010) Literature review and discussion on human thermal adaptation in built environment. J HV&AC 40(9):38–48
105. Karjalainen S (2012) Thermal comfort and gender: a literature review. Indoor Air 22(2):96–109
106. Indraganti M, Rao K (2010) Effect of age, gender, economic group and tenure on thermal comfort: a field study in residential buildings in hot and dry climate with seasonal variations. Energy Build 42(3):273–281
107. Schiavon S, Lee K (2013) Dynamic predictive clothing insulation models based on outdoor air and indoor operative temperatures. Build Environ 59:250–260
108. Zhai Y, Elsworth C, Arens E, Zhang H, Zhang Y, Zhao L (2015) Using air movement for comfort during moederate exercise. Build Environ 94(1):344–352
109. Humphreys MA, Nicol JF (1998) Understanding the adaptive approach to thermal comfort. ASHRAE Trans 104(1):991–1004
110. Kaczmarczyk J, Melikov A, Sliva D (2010) Effect of warm air supplied facially on occupants' comfort. Build Environ 45:848–855
111. Zhai Y, Zhang H, Zhang Y, Pasut W, Arens E, Lin Q (2013) Comfort under personally controlled air movement in warm and humid environments. Build Environ 65:109–117
112. Huang L, Ouyang Q, Zhu Y, Jiang L (2013) A study about the demand for air movement in warm environment. Build Environ 61:27–33

113. Candido C, de Dear R, Lamberts R, Bittencourt L (2010) Air movement acceptability limits and thermal comfort in Brazil's hot humid climate zone. Build Environ 45(1):222–229
114. Zhang H, Arens E, Kim DE, Buchberger E, Bauman F, Huizenga C (2010) Comfort, perceived air quality, and work performance in a low-power task-ambient conditioning system. Build Environ 45:29–39
115. Zhang H, Arens E, Pasut W (2011) Air temperature thresholds for indoor comfort and perceived air quality. Build Res Inf 39:134–144
116. Bauman F, Carter T, Baughman A, Arens E (1998) Field study of the impact of a desktop task/ambient conditioning system in office buildings. ASHRAE Trans 104(98):1153–1171
117. Zhang H, Arens E, Zhai Y (2015) A review of the corrective power of personal comfort systems in non-neutral ambient environments. Build Environ 91:15–41
118. van Marken Lichtenbelt W, Kingma B, van der Lans A, Schellen L (2014) Cold exposure—an approach to increasing energy expenditure in humans. Trends Endocrinol Metab 25(4):165–167
119. Hanssen M, Hoeks J, Brans B, van der Lans A et al (2015) Short-term cold acclimation improves insulin sensitivity in patients with type 2 diabetes mellitus. Nat Med 21(8):863–865
120. Pallubinsky H, Schellen L, Kingma B, van Marken Lichtenbelt W (2015) Human thermoneutral zone and thermal comfort zone: effects of mild heat acclimation. Extreme Physiol Med 4(1):1
121. Yu J (2012) Studies on the effects of physiological acclimation on thermal responses of people accustomed to different thermal indoor environments. Donghua University, Shanghai
122. Fanger PO, Toftum J (2002) Extension of the PMV model to non-air-conditioned buildings in warm climates. Energy Build 34(6):533–536
123. Wong NH, Khoo SS (2003) Thermal comfort in classrooms in the tropics. Energy Build 35(4):337–351
124. Zhang G, Zheng C, Yang W et al (2007) Thermal comfort investigation of naturally ventilated classrooms in a subtropical region. Indoor Built Environ 16(2):148–158
125. Zhou X, Zhu Y, Qin O et al (2014) Experimental study of the influence of anticipated control on human thermal sensation and thermal comfort. Indoor Air 24(2):171–177
126. Leaman A, Bordass B (2007) Are users more tolerant of 'green' buildings? Build Res Inf 35(6):662–673
127. Leaman A, Bordass B (2001) Assessing building performance in use 4: the probe occupant surveys and their implications. Build Res Inf 29(2):129–143
128. Nicol F, Humphreys M (2010) Derivation of the adaptive equations for thermal comfort in free-running buildings in European standard EN15251. Build Environ 45(1):11–17
129. Cartney K, Nicol F (2002) Developing and adaptive control algorithm for Europe. Energy Build 34(6):623–635
130. Toftum J (2012) Indoor Climate Survey at Espergaerde Gymnasium. Master thesis. Technical University of Denmark
131. Singh M, Mahapatra S, Teller J (2014) Relation between indoor thermal environment and renovation in Liege residential buildings. Therm Sci 18(3):889–902
132. Kim H (2012) Methodology for rating a building's overall performance based on the ASHRAE/CIBSE/USGBC performance measurement protocols for commercial buildings. Ph.D. thesis. Texas A&M University
133. Langevin J, Gurian P, Wen J (2015) Tracking the human-building interaction: a longitudinal field study of occupant behavior in air-conditioned offices. J Environ Psychol 42:94–115
134. Drake S, de Dear R, Alessi A, Deuble M (2010) Occupant comfort in naturally ventilated and mixed-mode spaces within air-conditioned offices. Archit Sci Rev 53(3):297–306
135. Vecchi R, Candido C, Lamberts R (2012) Thermal history and its influence on occupants' thermal acceptability and cooling preferences in warm-humid climates: a new desire for comfort? In: Proceedings of 7th Windsor conference: the changing context of comfort in an unpredictable world. Windsor, UK
136. Romero R, Bojorquez G, Corral M, Gallegos R (2013) Energy and the occupant's thermal perception of low-income dwellings in hot-dry climate: Mexicali, Mexico. Renew Energy 49:267–270

137. Rijal H, Humphreys M, Nicol F (2015) Adaptive thermal comfort in Japanese houses during the summer season: behavioral Adaptation and the effect of humidity. Buildings 5(3):1037–1054
138. Indraganti M, Ooka R, Rijal H, Brager G (2014) Adaptive model of thermal comfort for offices in hot and humid climates of China. Build Environ 74:39–53
139. Andamon M (2006) Thermal comfort and building energy consumption in the Philippine context. In: The 23rd conference on passive and low energy architecture, Geneva, Switzerland
140. Heidari S, Sharples S (2002) A comparative analysis of short-term and long-term thermal comfort surveys in Iran. Energy Build 34(6):607–614
141. Oseland N (1998) Acceptable temperature ranges in naturally ventilated and air-conditioned offices. ASHTAE Trans 104:1018
142. Bouden C, Ghrab N (2005) An adaptive thermal comfort model for the Ynisian context: a field study results. Energy Build 37(9):952–963
143. Yu J, Cao G, Cui W, Ouyang Q, Zhu Y (2013) People who live in a cold climate: thermal adaptation differences based on availability of heating. Indoor Air 23(4):303–310
144. Yang L, Yan H, Xu Y, Lam J (2013) Residential thermal environment in cold climates at high altitudes and building energy use implications. Energy Build 62:139–145
145. Yao R, Li B, Liu J (2009) A theoretical adaptive model of thermal comfort—adaptive predicted mean vote (aPMV). Build Environ 44(10):2089–2096
146. Yao R, Liu J, Li B (2010) Occupants' adaptive responses and perception of thermal environment in naturally conditioned university classrooms. Appl Energy 87(3):1015–1022
147. Liu H, Zheng W, Li B et al (2011) Behavioral adaptation of indoor thermal environment in hot-summer and cold-winter zone. J Central South Univ 42(6):1805–1812
148. Wang Z, Li A, He Y et al (2012) Human thermal comfort and thermal adaptability in Harbin. J Harbin Inst Technol 44(8):48–52
149. Ning H, Wang Z, Zhang X, Ji Y (2016) Adaptive thermal comfort in university dormitories in the severe cold area of China. Build Environ 99:161–169
150. Zhang Y (2011) Review and discussion on thermal adaptation research approaches in built environment. J HV&AC 41(2):9–17
151. Zhang Y, Chen H, Meng Q (2013) Thermal comfort in buildings with split air-conditioners in hot-humid area of China. Build Environ 64:213–224
152. Zhang Y, Wang J, Chen H, Zhang J, Meng Q (2010) Thermal comfort in naturally ventilated buildings in hot-humid area of China. Build Environ 45(11):2652–2570
153. Zhang Y, Chen H, Wang J, Meng Q (2016) Thermal comfort of people in the hot and humid area of China—impacts of season, climate, and thermal history. Indoor Air 26(5):820–830
154. Mavrogianni A, Johnson F, Ucci M et al (2013) Historic variations in winter indoor domestic temperatures and potential implications for body weight gain. Indoor and Built Environment. 22(2):360–375

Chapter 2
Indoor Climate Experience and Thermal Comfort Expectation in Buildings

2.1 Background

2.1.1 Air-Conditioning and Indoor Thermal Environment

The mutually dependent relationship between users' comfort expectation and HVAC technological development in western countries was noticed by Reyner Banham in the 1960s [1]. He stated that the most significant architecture changes caused by artificial environmental control development involve the manifestation of expectation changes, including both changes in user's needs and changes in methods of servicing users' needs. In the 1990s, Kempton [2] mentioned that the relationship between comfort expectations and air-conditioning, at least in the US, became so strong that it was often referred as an "addiction" similar to clinicians with respect to drugs. At present, the escalating AC market penetration in developing countries indicate that this kind of "addiction" not only happened in the US but also is happening in global demographic superpowers like Mexico, Brazil, India, and China.

The excessive need for comfort brought about indoor thermal environment homogenization, compressing the wide "natural range" of indoor temperature to narrow thermal neutralities [3]. The pervasiveness of HVAC throughout homes, workplaces, and transport makes our daily thermal exposures ever closer to theoretically ideal conditions: warm indoor environments being cooled while cool conditions being heated. The benefits of being able to mechanically heat and cool our buildings to eliminate uncomfortable conditions are widely accepted, of course. The question becomes—what is "ideal", and how narrowly or universally are ideal conditions defined? Taking wintertime indoor residential temperature as an example, Fig. 2.1 shows its temporal shifting trends in countries like US, UK, northern Japan, and China. It shows that typical indoor winter temperatures in these countries are

M. Luo, *The Dynamics and Mechanism of Human Thermal Adaptation in Building Environment*, Springer Theses,
https://doi.org/10.1007/978-981-15-1165-3_2

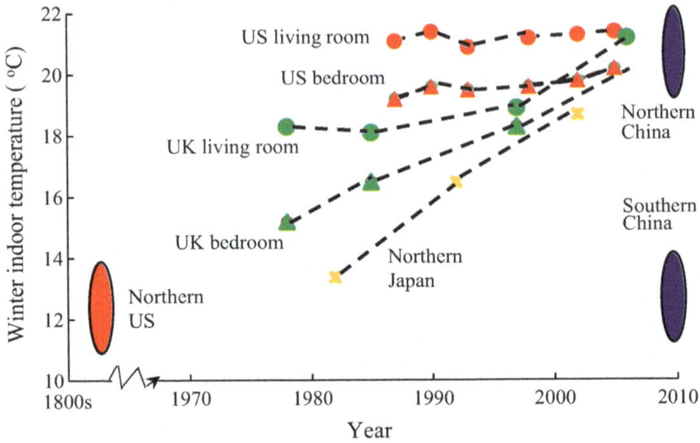

Fig. 2.1 Historical trends of wintertime residential temperatures (US 1988–2008 and UK Sources are based on the review work of [6]; US 1800s [7]; Northern Japan [8]; China [9, 10])

narrowing over time, converging to some neutral temperature of about 20 °C. Some researcher even named this trend as 'homogenization of built environment' [4] and 'comfort capsules' [5].

2.1.2 *Thermal Comfort Expectation*

Under the contexts of "indoor climate homogenization", researchers are questioning whether such improvement in the indoor thermal environment has produced a proportionate increment in building occupants' thermal satisfaction. Analyzing the ASHRAE database field study data, Arens et al. [11] showed that tight "class A" temperature control does not necessarily translate into higher thermal comfort evaluation compared with less tight "class C" control. About 20% of building occupants reported their thermal environment uncomfortable no matter how tightly their thermal environment was controlled. Usually, such thermal discomfort complaints were explained by the interpersonal difference in thermal comfort demand and local thermal discomfort. But it leads one to question what is really meant by "classes" of environments. While early definitions tied the classes to the tightness or narrowness of the temperature set-points, perhaps it should be more related to the degree of personal control occupants have. However, the findings suggest another question we might ask. Will people living in what might be considered a "high quality" indoor environment (which, admittedly, can be defined in various ways) have their expectations raised, thereby leading to higher demands or higher frequency of complaints, while their counterparts with lower quality (or perhaps simply "more variable") indoor environments unconsciously lower their thermal demand by adaptation? This question goes very core to the notion of thermal comfort expectation, which was firstly acknowledged in McIntyre's work [12]. In the 1990s, Brager and de Dear [13, 14] introduced the expectation concept into thermal comfort research field as a psychological adapting approach explaining why people are more accepting of the wider ranges of thermal conditions in naturally ventilated building environments. In 2002, Fanger and Toftum [15] induced the "expectation" as the seventh parameter in their Predicted Mean Vote (PMV) equation to extend PMV application in non-air-conditioned buildings in warm climates.

Although few studies have investigated the issue of expectation directly, some studies begin to allude to the potential effects. For example, a study by Nicol and Humphreys [16] found the comfortable indoor temperatures in heated or cooled office buildings in the 1990s were much more tightly clustered than those observed in 1970s. Rajasekar [17] in their field study found that people in the hot humid climatic region of India had a thermal expectancy factor of 0.7. Andamon et al. [18] studied how the modern transformations, first colonization, and then the introduction of air conditioning changed the Philippine people's understanding of normal and ordinary conditions of comfort. One of their conclusions is that the social understanding of comfort and its transition is malleable. Amin et al. [19] investigated the influence of occupants' thermal history on the use of controls and indoor temperature preference in a newly built student hall in the UK. Their result indicates the average indoor temperature of residents from warm climates was 2.3 °C higher than those from cool climates. Wang et al. [20] showed that people in the severe cold area of China tended to get used to the indoor temperature gradually if their apartments were overheated.

While these studies hint at the notion of thermal comfort expectations, they fail to offer solid evidence supporting the impacts of expectation, which is a complex combination with many confounding factors such as cultural difference and economic level. Direct empirical data needed to elaborate the underlying principles and dynamics of comfort expectation are yet to be produced. Given that the global demographic superpowers like India and China are embarking upon tremendous construction boom, the significance of thermal comfort expectation and their potential impacts on building design and operation cannot be overstated.

2.1.3 *Objectives of This Study*

In a previous paper [21], we revisited the notion of comfort expectation and designed an intriguing field-based experiment in China to explore the inherent principle of thermal comfort expectation. The current study further analyzed the experimental results in a more thorough way to answer the following questions: (1) what is the relationship between people's indoor thermal experiences and their thermal comfort expectations? (2) how can one describe the changes in thermal comfort expectation? (3) what are the dynamic characteristics of thermal comfort expectation?

2.2 Method

2.2.1 Wintertime Indoor Temperatures in China

Since the 1950s, China has been implementing a space heating policy setting the Huai River as the boundary of centralized district heating between southern and northern cities. According to this policy, centralized district heating networks were only established in northern cities, covering most of the Severe Cold and the Cold zones. Meanwhile, district heating systems were prohibited in southern China which includes the Hot Summer and Cold Winter zone, the Temperate zone, and the Hot Summer and Warm Winter zone (shown in Fig. 2.2).

As a legacy of the Huai River policy, recorded wintertime indoor temperatures in northern cities have been much warmer than those in southern cities. This generalization is supported by many field studies with thermal environment measurements, as partially summarized in Table 2.1. One typical example is the large-scale survey done by Yoshino et al. [9], focusing on residential indoor thermal environments in major Chinese cities in different climate conditions. In this study, surveys were simultaneously conducted in what will be characterized as northern China (Harbin, Urumqi, Beijing, Xi'an), and in southern China (Shanghai, Chongqing, Changsha, Kunming, Hong Kong), and others. The measured results collectively support the finding that, in cold northern cities with centralized heating networks, wintertime indoor temperatures were relatively stable around 20 °C. At the same time, indoor temperatures in southern cities which lacked district heating systems, especially those in the Hot

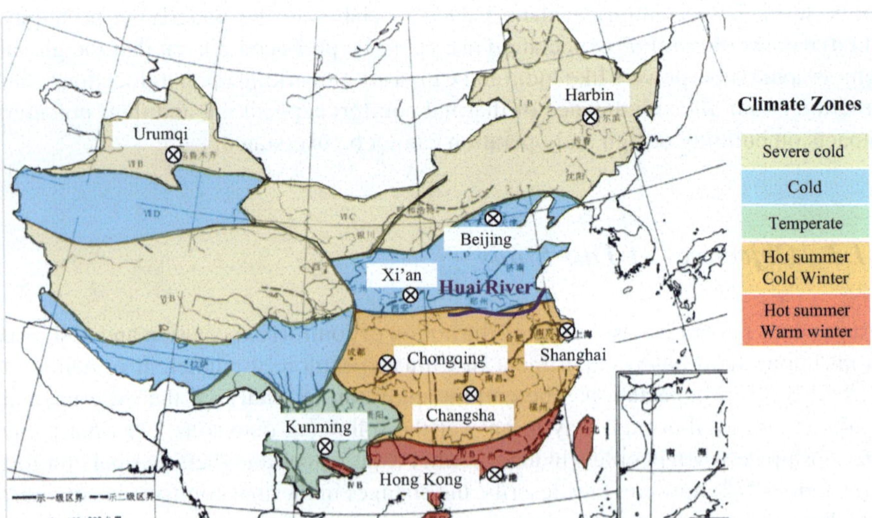

Fig. 2.2 Heating zones in China

Table 2.1 Measured residential wintertime indoor temperature and humidity in typical cities

Climate zones	Space heating days	Cities	Indoor temperature (°C)	Indoor relative humidity (%)	References
Northern China					
Severe cold	150–200	Harbin	20–24	20–35	[9, 22]
		Urumqi	20–24	20–35	[9]
Cold	90–150	Beijing	19–24	25–40	[9, 23]
		Xi'an	19–24	20–40	[9]
Southern China					
Hot summer Cold winter	–	Shanghai	8–14	65–80	[9, 23]
		Chongqing	8–14	70–85	[9, 24]
		Changsha	8–13	70–85	[9, 25]
Temperate	–	Kunming	13–17	55–75	[9]
Hot summer warm winter	–	Hong Kong	16–23	60–80	[9]

Summer and Cold Winter zone, were typically as cold as 8–14 °C. A similar generalization can be observed in the humidity measurements as well, where the relative humidity in northern cities was about 30%, while that in southern cities could be as humid as 70%.

2.2.2 Subject Groups

The indoor temperature difference between northern and southern China created natural conditions for an intriguing experiment. In addition to looking at differences simply between people who lived in these different regions, we also wanted to look at the differences in people who grew up in that region versus those who moved between regions. With this in mind, the research team defined four subject groups: (1) subjects who had always lived in northern China (N-N); (2) subjects who had moved from southern China to northern China (S-N); (3) subjects who had always lived in southern China (S-S); and (4) subjects who had moved from northern China to southern China (N-S). As subjects' indoor thermal experience is closely related to their geographical location, it is important to identify where they live now and where they came from. In this study, northern subjects mainly came from or lived in cities like Beijing, Tianjin, Xi'an and Harbin, southern subjects mainly came from or lived in cities around Yangzi River Basin such as Shanghai, Chongqing, Wuhan, and Changsha.

Although there were clear differences in the availability of centralized district heating in the two regions, we also wanted to examine the influence of localized space heating on indoor temperature. Figure 2.3 shows significantly different patterns

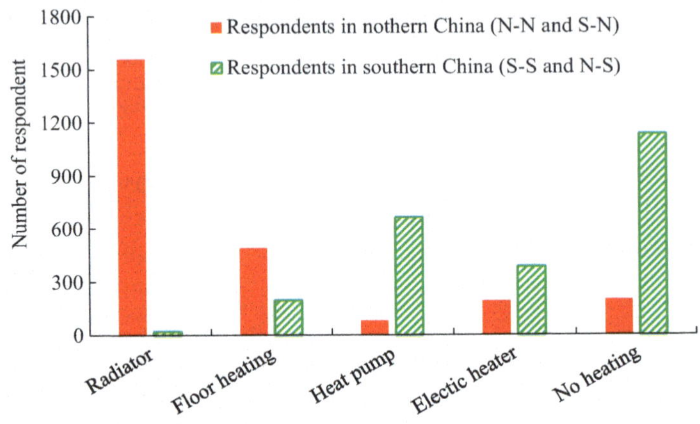

Fig. 2.3 Statistics of heating facilities

in what kinds of local heating terminals were utilized by respondents in the two regions. In the colder northern China regions, 81% of respondents used radiator heating or floor heating, compared to southern China where 91% of respondents used air source heat pumps, electronic heater or just no heating. To ensure a cleaner separation between the regional differences in both wintertime indoor temperatures, and the types of heating systems that were used, the following analysis excluded subjects who lived in southern China but utilized the systems more common to northern China—floor heating and radiator heating, and also excluded those living in northern China but without the availability of district heating facilities. In total, 4411 respondents among the original 5168 were selected for analysis. As shown in Table 2.2, similar numbers of male and female subjects were recruited, and over 80% of each subject group were young college students aged 20–30 years old, aiming to minimize the confounding effects of age.

2.2.3 *Survey Design*

The online survey was administered during two periods: December 2014 through February 2015, and December 2015 through February 2016. These two periods were typically the coldest time during a whole year in China. In this way, the difference in indoor temperature between southern and northern regions can be guaranteed because southern China homes lacked continuous space heating while northern area homes had centralized space heating. Subjects were asked to evaluate the general wintertime indoor thermal environment of their hometown based on their memory and also what they think about indoor climates in general of their current home. The word "general" refers to the overall impression during the wintertime instead of "right here right now" surveys that are often used in thermal comfort field studies.

Table 2.2 Subject group information

Group	Winter indoor thermal experience	Number of subjects			Age				
		Total	Male	Female	<20	20–30	30–40	40–50	>50
N-N	Always lived in northern China	988	430	558	73	498	128	177	112
S-N	Moved from southern to northern China	1138	628	510	169	912	14	22	21
S-S	Always lived in southern China	1616	644	972	124	1301	65	54	72
N-S	Moved from northern to southern China	669	393	276	21	562	40	26	20

Table 2.3 Question scales

	Thermal sensation	Thermal comfort	Humidity sensation	Humidity comfort	Overall acceptance
3	Hot	Very comfortable	Very dry	Very comfortable	Very acceptable
2	Warm	Comfortable	Dry	Comfortable	Acceptable
1	Slightly warm	Just comfortable	Slightly dry	Just comfortable	Just acceptable
0	Neutral	Unclear	Neutral	Unclear	Neutral
−1	Slightly cool	Just uncomfortable	Slightly humid	Just uncomfortable	Just unacceptable
−2	Cool	Uncomfortable	Humid	Uncomfortable	Unacceptable
−3	Cold	Very uncomfortable	Very humid	Very uncomfortable	Very unacceptable

For example, subjects in the N-S group were asked to assess both North and South areas; for the North, it's their memory before they migrated, and for the South, it's their general impression on the current home. The survey was focused on residential buildings such as a house, apartment or dormitory. The survey questions all used seven-point scales and asked about thermal sensation vote, thermal comfort vote, humidity sensation, humidity comfort, and overall acceptance. The words used in each of these seven-point scales are depicted in Table 2.3.

2.2.4 Data Analysis

Calculations were made for the descriptive statistics (mean, standard deviation) and the differences between means of thermal sensation vote and thermal comfort vote. Two sample *T*-tests with equal variances were applied to verify the significance of the differences among groups. For all tests, the results were considered statistically significant when $p \leq 0.05$. The interpretation of the outcome was as follows: $p \leq 0.001$ means highly significant, $0.001 < p \leq 0.01$ means significant, $0.01 < p \leq 0.05$ means weakly significant, and $p > 0.05$ means not significant. Due to the size of the samples possibly confounding the *T*-test result, for each comparison, we calculated the effect size (Cohen's d) to quantify the impact of sample sizes on the statistically significant differences. The interpretation of the outcome followed the thresholds provided by Sawilowsky [26]: d = 0.01 means very small difference, d = 0.2 means small difference, d = 0.5 means medium difference, d = 0.8 means large difference, d = 1.2 means very large difference, and d = 2.0 means a huge difference. All statistical analysis was performed using Excel version 2016.

2.3 Results

2.3.1 *Indoor Thermal History and Thermal Adaptation*

Table 2.4 summarizes each group's evaluations of the general indoor thermal environments of residential buildings in southern and northern China. Two initial major findings can be drawn from it. Firstly, the N-N group's thermal sensation vote (TSV) was dominated by 'neutral' and 'warm', while the majority of S-S group felt 'cool' and 'slightly cold'. This is consistent with the earlier mentioned thermal environment generalization between northern and southern China. But what's interesting is that although these two groups had different thermal sensations, they didn't show a significant difference in their thermal comfort vote (TCV). Feedback from both groups showed that a significant majority of TCV fell in the "comfortable" range. Secondly, the two migrating groups (N-S and S-N) had similar TSV and TCV evaluations with N-N group on northern China's wintertime indoor temperature, but significantly worse TSV and TCV (felt colder and more uncomfortable) than S-S group on southern China's indoor temperature.

Figure 2.4a shows the overall distribution of thermal sensation votes between different thermal history groups, with the upper bars representing people who had moved between regions, and the lower bars representing people who remained in the same region. There were stark differences in people's thermal sensation votes when asked about the two regions. The majority of occupants' impression of the northern areas were either neutral or warm (and of those who expressed cold sensations, there were more from people who always lived in that region, compared to those who moved either from or to the region). In comparison, over 70% of thermal sensations about southern China were dominated by cold side votes, with the strongest impressions (over 90% cold) from people who no longer lived there are were voting about their memory of the experiences.

Similarly, Fig. 2.4b compares the overall distribution of comfort perceptions between different groups. Despite the widely discrepant indoor temperatures and thermal sensations (Fig. 2.4a), the N-N and S-S groups had close thermal comfort evaluations; both groups had a majority of positive evaluations, accounting for 71 and 60% respectively. But the two migrating groups (N-S and S-N) expressed completely different comfort perceptions about northern and southern China, regardless of in which direction they migrated. They both gave overwhelmingly positive feedbacks on northern China's wintertime indoor thermal environment (65–83% comfortable) while giving more negative evaluations of the southern area (only 41–52% comfortable).

To better understand the complicated relationships between each groups' indoor thermal history and their corresponding thermal comfort perception, an index named 'comfort score' was defined as Eq. 2.1. This score would return a value between +1.0 if all of the votes were in the comfortable range, and −1.0 if all the votes were uncomfortable. To do the calculation, the votes from 'just uncomfortable' to 'very uncomfortable' votes were categorized as the '***uncomfortable side***',

Table 2.4 Different subject groups' indoor thermal environment evaluation

Subject groups		N-N	S-S	N-S		S-N	
Region		North	South	North	South	North	South
Thermal sensation vote (TSV)	Cold	14	380	3	154	1	266
	Cool	57	748	25	243	17	543
	Slightly cool	343	1252	42	107	54	257
	Neutral	1124	745	415	118	677	58
	Slightly warm	249	96	155	35	200	13
	Warm	151	9	25	11	151	1
	Hot	38	8	4	0	38	0
Mean		0.08	−1.17	0.17	−1.49	0.46	−1.87
SD		0.98	1.03	0.81	1.26	0.95	0.87
T-test with equal variances	Compare with N-N group	–	<0.001	o	<0.001	o	<0.001
	Compare with S-S group	<0.001	–	<0.001	<0.001	<0.001	<0.001
Effect size (d)	N-N as control group	–	**1.28**	0.09	**1.62**	0.39	**2.00**
	S-S as control group	**1.22**	–	**1.31**	0.32	**1.59**	0.68
Thermal comfort vote (TCV)	Very uncomfortable	24	73	3	43	7	44
	Uncomfortable	139	334	16	118	42	248
	Just uncomfortable	309	693	44	119	141	342
	Unclear	204	374	46	56	112	114
	Just comfortable	798	1385	304	182	489	298
	Comfortable	304	286	193	38	321	83
	Very comfortable	198	87	63	13	26	9
Mean		0.68	0.20	1.19	−0.32	0.85	−0.42
SD		1.42	1.35	1.12	1.59	1.17	1.40
T-test with equal variances	Compare with N-N group	–	0.041	o	<0.001	o	<0.001
	Compare with S-S group	0.041	–	o	<0.001	<0.001	<0.001
Effect size (d)	N-N as control group	–	0.34	0.36	**0.70**	0.12	**0.78**
	S-S as control group	0.36	–	**0.73**	0.38	0.48	0.46

Note T-test with no significant difference was labeled as 'o'. Effect size with Cohen's d larger than 0.7 was marked as a bold font. As each group were asked to give evaluations to their hometown and their current abode place, N-S, and S-N groups had evaluations on both North and South, and N-N and S-S groups had two times of voting number than their example number

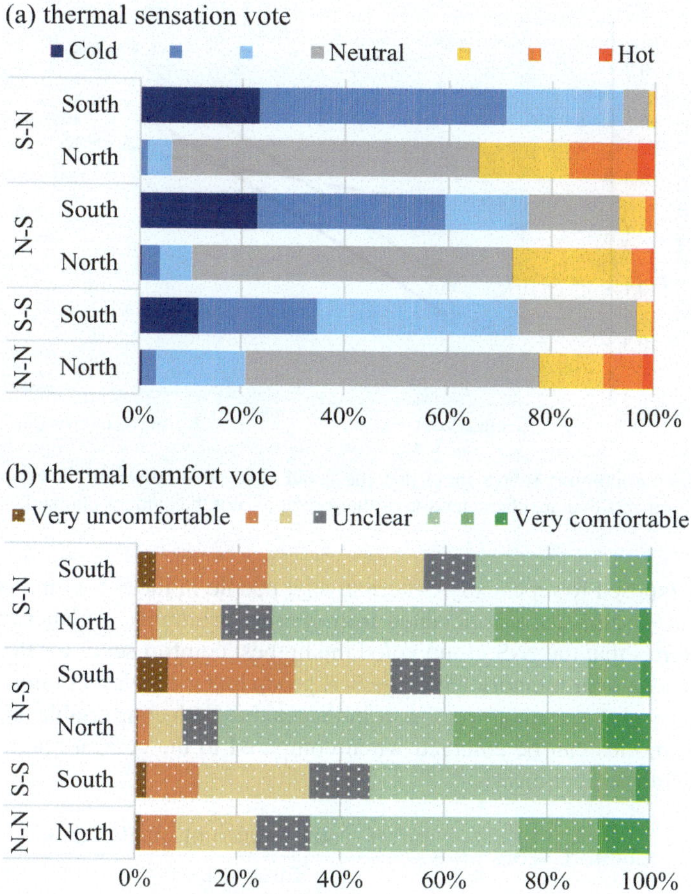

Fig. 2.4 Different groups' thermal perception

while votes from 'just comfortable' to 'very comfortable' were grouped into the '***comfortable side***'. As for the 'unclear' votes, it was split into half and half—50% were regarded as comfortable and the rest 50% were grouped into the uncomfortable side.

Figure 2.5 compares the different groups' comfort scores and shows that people living in colder indoor thermal environments (group S-S) had quite similar comfort scores with those living in neutral-to-warm indoor conditions (group N-N), despite the fact that they experienced physically different indoor thermal environment. But for people who migrated and therefore experienced both non-heated and heated indoor thermal environments (group N-S and S-N), they reported higher comfortable scores for the northern region. This suggests that people's understanding of thermal comfort is malleable, depending on their previous thermal experience. People in the N-N group had been shown to adapt to thermal neutrality if that was what

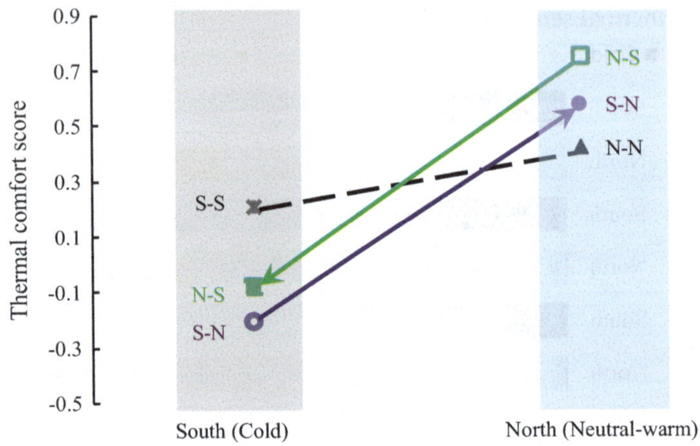

Fig. 2.5 Thermal comfort scores (note that the "cold" and "neutral-warm" labels refer to the *indoor* environments that are characteristic of these regions, not the outdoor climate)

they were exposed to repeatedly. Alternatively, people in the S-S group could also adapt to indoor climates that deviated from thermal neutrality. Another interesting phenomenon is that the N-S group voted the highest comfort score for the northern region based on their memory while the S-N group had the lowest comfort score for their memory of the southern region. It indicates that both comfortable and uncomfortable memories can be enlarged when compared to the current comfortable or uncomfortable perceptions.

$$\text{Comfort score} = \frac{\text{comfort votes} - \text{uncomfortable votes}}{\text{Total votes}} \tag{2.1}$$

2.3.2 *Dynamic Changes in Thermal Adaptation*

The above result shows a mutually dependent relationship between the indoor thermal environment and occupants' understanding of thermal comfort, which is reflective of thermal adaptation. But how do repeated exposures to different indoor temperatures shape our expectations of the thermal environment? To study this, we further subdivided the migrating groups (S-N and N-S) on the basis of duration in the new region following migration. The number of samples for each subgroup is shown in Table 2.5.

Figure 2.6a, b show the trend of comfort evaluations of the N-S and S-N subgroups, respectively, based on how long they had lived in the new region. In Fig. 2.6a, it can be seen that comfort evaluations of the migrating northerners started out at the lowest (30.4%), and then improved gradually as people lived in the south longer. People

Table 2.5 Subgroup samples based on time in a new location following migration

	Less than 1 year	1–2 year	2–3 year	3–4 year	More than 4 years	Total
N-S	106	93	111	144	216	669
S-N	228	132	152	324	302	1138

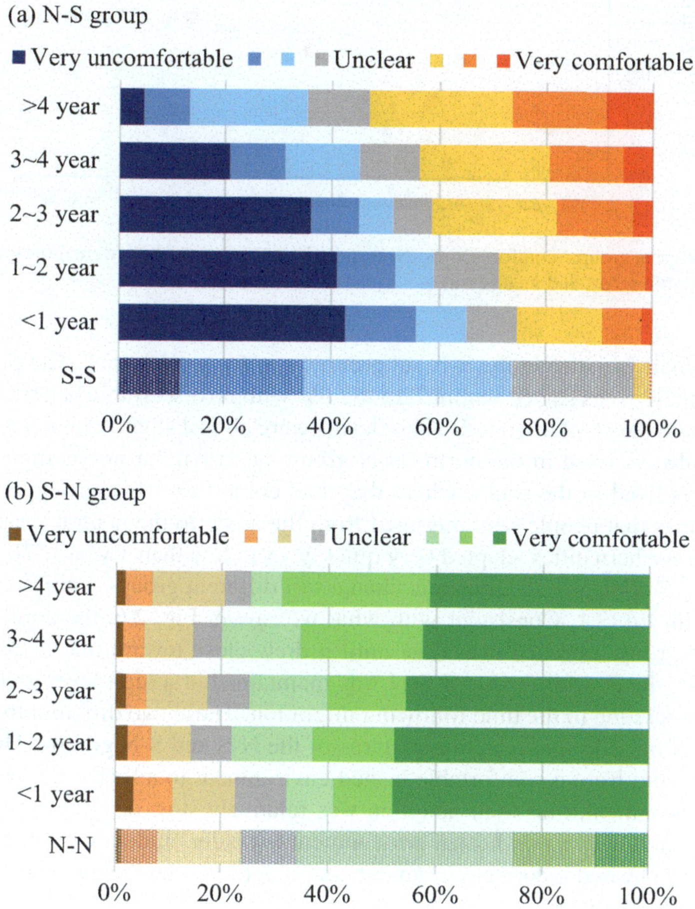

Fig. 2.6 The changes in migrating groups' comfort evaluation

who had lived there for at least four years had essentially the same responses (58.9% comfortable) as those who had always lived in the south (S-S group), as shown in the rightmost column. This can be regarded as an indicator that the migrated northerners required time to adapt to the non-heated southern environments, adapting more in each year after they migrated. This same pattern did not hold true, however, for people who migrated from the south to the colder north, as shown in Fig. 2.6b where there

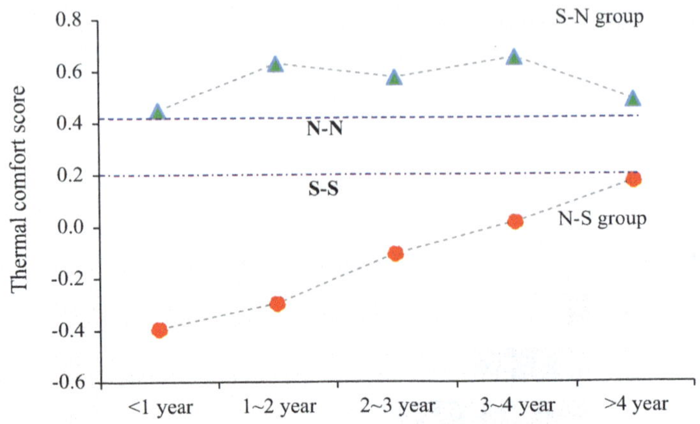

Fig. 2.7 Different groups' comfort scores (Note, positive values mean there were more comfortable as uncomfortable votes, and vice versa)

is a very small variance in the S-N subgroup's comfort evaluations. The proportion of comfortable votes stayed within 72.5–82.6% with no discernible pattern based on how long ago they had migrated. These levels were all just slightly higher than those who had always lived in the north (N-N group, at 71%), but not as high as those who always lived in the south, where they had come from (S-S group, at 60.2%). This suggests that people who migrated from the south to the neutral-warm heated indoors in northern cities adapted very quickly, even less than 1 year [27].

Figure 2.7 compares the dynamic changes of different groups' comfort score, as expressed in Eq. 2.1. Consistent with what we saw in Fig. 2.6, the comfort score of N-S subgroups climbed over time until it drew close toward that of S-S group. Meanwhile, the comfort score of S-N was maintained at a high level with a small variance in relation to the time following migration. These two different trends may reflective of the different adaption patterns of the N-S and S-N groups. The former group migrated from a comfortable heated environment to a less comfortable non-heated environment, thus their adaption was relatively slow or reluctant; the latter group migrated from a non-heated environment to colder outdoor climate but with a neutral-warm heated indoor environment, and it appears that the adaptation process was relatively quick.

The above comfort score analysis helped to understand the dynamics of thermal adaptation in this region to some extent, but this method has limitations as well. For example, it didn't reveal any nuances in the adaptation process of the S-N group. In order to compare different groups' adaptation process in more detail, we proposed a new metric, which was called the 'demand factor' (DF) index, to describe people' thermal comfort demand for the thermal environment. The basic logic of the DF index is as follows: if a group of people has relatively high requirements (or expectations) of the indoor thermal environment, and lower tolerance of uncomfortable conditions (i.e., deviations from thermal neutrality), the combination of these factors would lead to a bigger value of DF. On the contrary, if a group of people has relatively low requirements of the thermal environment, and higher tolerance of uncomfortable conditions, this would have a smaller value of DF. To quantify this, we define DF as shown in Eq. 2.2.

$$\text{Demand factor}_i = \frac{\sum_{\text{cold}}^{\text{hot}} \text{percentage of uncomfortable}_i}{\sum_{\text{cold}}^{\text{hot}} \text{percentage of uncomfortable}_{\text{baseline}}} \tag{2.2}$$

where 'i' means group$_i$, '$\sum_{\text{cold}}^{\text{hot}}$ percentage of uncomfortable$_i$' means the sum of group$_i$'s uncomfortable percentages across the range of cold-to-hot thermal sensation votes, '$\sum_{\text{cold}}^{\text{hot}}$ percentage of uncomfortable$_{\text{baseline}}$' means the sum of group$_{\text{baseline}}$'s uncomfortable percentages across the range of thermal sensation votes. It should be noted that the selection of the baseline, or reference, group is flexible. It can be a real group of people, or a hypothetical group determined using the PMV-PPD model. In this study, we selected the average of all the respondents as the baseline group, which includes S-S, N-N, S-N and N-S groups. When DF is >1, this means that group$_i$ had higher levels of discomfort, or a higher demand or expectations, compared to the baseline group, and vice versa when DF is <1.

Table 2.6 shows the relationship between uncomfortable votes (not including neutrality) and thermal sensation votes, and also the DF calculations, for the groups who have lived in each region all their lives. Within each thermal sensation category, the N-N group had higher uncomfortable votes than S-S group, which indicates that the N-N subjects had a higher demand for the thermal environment and were less tolerant of conditions that deviated from thermal neutrality. This is also revealed as a higher demand factor for the N-N group. According to the definition of Eq. 2.2 and the specific values listed in Table 2.6, the demand factor can be calculated as 0.8 for S-S group and 1.13 for N-N group.

The notion of demand factor can be extended to N-S and S-N subgroups. By comparing different groups' demand factor for the time since migration, Fig. 2.8 shows the dynamic changes of each group's thermal adaptation. It's interesting to see that the two migrated groups, N-S and S-N, exhibited different patterns in their demand factor over time, which is reflective of the process of adaptation. The demand factor of the N-S group decreased gradually (implying a slow rate of adaptation). This suggests that even after more than 4 years, they still didn't fully accept the cold indoor thermal environment in southern China. In comparison, the demand factor of the S-N group increased rapidly in the first year and then maintained at a certain level, which means they adapted quickly to the northern China indoor environments, and within a year were responding similarly to people who had lived in northern China their whole lives.

Figure 2.8 also suggests that it may be quicker to develop higher expectations (i.e., have a higher demand factor) while it is slower to lower one's demand factor (i.e., become accustomed to less comfortable conditions), which indicates that thermal adaptation in buildings is asymmetric. For example, the N-S group is those people

Table 2.6 Uncomfortable rate and demand factor for N-N and S-S groups

Uncomfortable % for each thermal sensation vote									
		Cold	Slightly cold	Cool	Neutral	Warm	Slightly hot	Hot	Demand factor
Baseline group	Percentage	77	54	41	14	26	60	85	–
	Number of vote	630	882	843	439	194	209	75	
N-N group	Percentage	95	65	44	15	28	68	90	1.13
	Number of vote	13	37	151	169	70	103	34	
S-S group	Percentage	64	52	41	14	21	35	60	0.80
	Number of vote	243	389	501	104	20	3	5	

Note that the percentage numbers in this table are uncomfortable percentage for each thermal sensation vote. That's why the summarization of the percentage row can exceed 100%

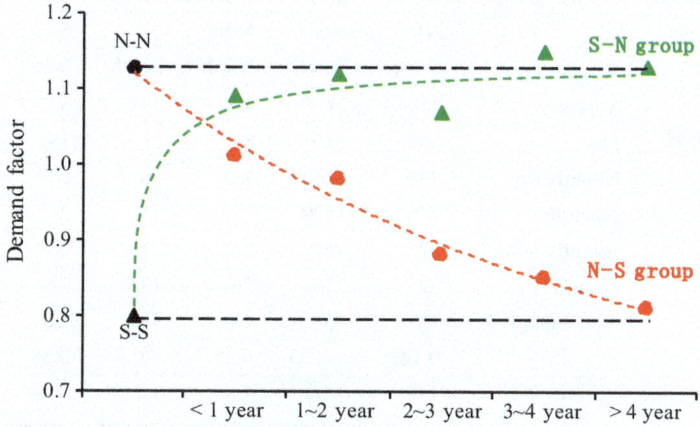

Fig. 2.8 Dynamics of thermal demand factor

who migrated from a relatively comfortable neutral-warm indoor thermal environment to a more uncomfortable cold indoor environment. The gradual decrease of their demand factor was slow; it took them more than 4 years to lower their thermal demand to a level close to people who lived in southern China all their life. On the other hand, the S-N group are people who moved from a relatively cold indoor environment to a more neutral indoor environment. Their demand factor increased rapidly in less than 1 year, and then maintained at a high level similar to that of N-N group. In short, it is slow and perhaps a reluctant process for occupants to move a neutral to a cold indoor climate, while it is quick and perhaps more delightful to get accustomed to the thermally neutral lifestyle.

2.3.3 *Dynamic Changes of Humidity Adaptation*

The above method can also be applied to analyze how people adapt to different humidity environments. Table 2.7 presents each group's indoor humidity environment evaluation in southern and northern China. The N-N group's humidity sensation vote was dominated by 'neutral' and 'slightly dry', while the majority of the S-S group felt 'neutral' and 'slightly humid'. This is consistent with the earlier mentioned generalization about the humidity environment between these two regions. But what's interesting is that although these two groups had different thermal sensations, they didn't show a significant difference in their humidity comfort vote. Feedback from both groups showed a significant majority of "just comfortable" vote. For the two migrating groups (N-S and S-N), the N-S group thought that northern China (with 'neutral-dry' humidity environments) was more comfortable than southern China,

Table 2.7 Different subject groups' indoor humidity environment evaluation

Subject groups		N-N	S-S	N-S		S-N	
Region		North	South	North	South	North	South
Humidity sensation vote	Very dry	113	24	42	13	52	5
	Dry	259	145	89	21	181	21
	Slightly dry	738	211	131	36	699	89
	Neutral	770	1579	292	176	233	375
	Slightly humid	52	964	69	215	36	461
	Humid	22	236	36	185	23	139
	Very humid	13	73	11	43	14	48
Mean		0.74	−0.33	0.39	−0.87	0.88	−0.65
SD		1.00	1.00	1.27	1.21	0.98	1.02
T-test with equal variances	Compare with N-N group	–	<0.001	<0.001	<0.001	<0.001	<0.001
	Compare with S-S group	<0.001	–	<0.001	<0.001	<0.001	<0.001
Effect size (d)	N-N as control group	–	**0.96**	0.33	**1.24**	0.14	**1.15**
	S-S as control group	**0.96**	–	0.67	0.45	**1.08**	0.31
Humidity comfort vote (TCV)	Very uncomfortable	56	86	34	135	71	45
	Uncomfortable	141	163	52	113	136	152
	Just uncomfortable	203	251	159	92	364	204
	Unclear	250	194	114	104	122	106
	Just comfortable	918	1627	139	122	274	303
	Comfortable	283	673	99	77	126	196
	Very comfortable	126	238	82	26	56	132
Mean		0.61	0.88	0.32	−0.55	−0.13	0.39
SD		1.37	1.32	1.66	1.82	1.56	1.71
T-test with equal variances	Compare with N-N group	–	<0.001	<0.001	<0.001	<0.001	<0.001
	Compare with S-S group	<0.001	–	<0.001	<0.001	<0.001	<0.001
Effect size (d)	N-N as control group	–	0.20	0.20	**0.74**	0.50	0.14
	S-S as control group	0.20	–	0.40	**0.94**	**0.70**	0.34

Note *T*-test with no significant difference was labeled as 'o'. Effect size with Cohen's d larger than 0.7 was marked as a bold font

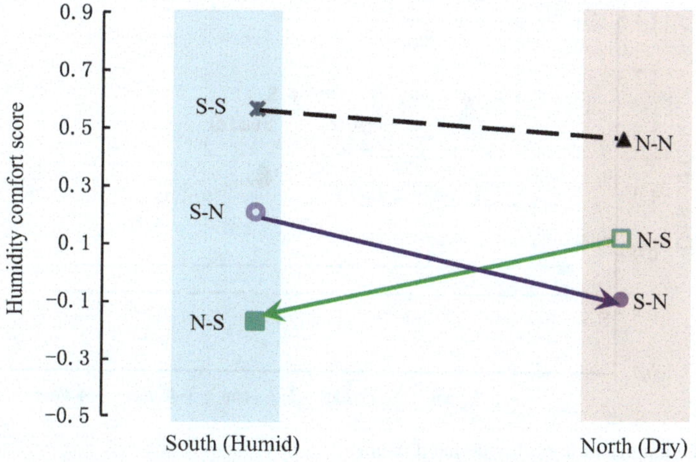

Fig. 2.9 Humidity comfort evaluations of the south and north indoor environments

while the S-N believed that the southern area (with 'neutral-humid' humidity environments) were more comfortable. In other words, they both perceived the humidity environments of their original locations to be more comfortable.

Figure 2.9 compares different groups' "humidity comfort scores", i.e., looking at the difference between comfortable and uncomfortable votes, again using Eq. 2.1 but using the humidity comfort votes shown in Table 2.7. As noted previously, the questions asked people to assess their perceptions of the indoor environment in each region—whether they lived there permanently or migrated from or to the region. It shows that the S-S group permanently living in relatively humid climates had a quite similar indoor humidity comfort score (0.57) compared to the N-N group living in dry climates (0.47), despite the fact that they experienced quite different indoor humidity (shown in Table 2.1). However, people who migrated and therefore experienced both humid (southern) and dry (northern) indoor climates expressed much more positive evaluations of their original living places (humidity comfort index > 0) compared to relatively more negative evaluations of their current living places (humidity comfort index < 0). This means that when you look at impressions of a given region, groups who migrated to that region always had a lower humidity comfort index than those who either permanently or originally lived there. This phenomenon suggests that humidity comfort perception is long-lasting and is closely related to the environments you were exposed to growing up.

By comparing Fig. 2.9 and Fig. 2.7, one can notice that people adapt to temperature and humidity environments in different ways. For thermal comfort assessments shown in Fig. 1.6, Chap. 1, the indoor environments in the non-heated South China area are cold, and receive a lower comfort score, while the heated North China regions have neutral-warm indoor conditions and received higher comfort scores. The S-N group migrated from lower quality to the higher quality environment in terms of indoor temperature assessments, and the adaptation process was relatively

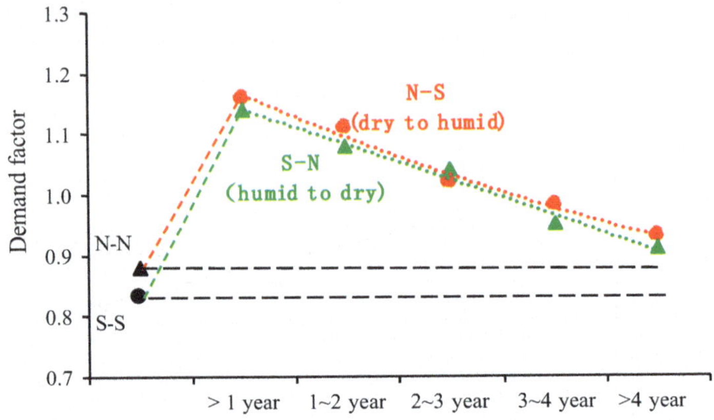

Fig. 2.10 Dynamics of humidity demand factor

quick; N-S group moved in the other direction, from a higher to lower quality indoor environment, and their adaptation is relatively slow. But for humidity environment shown in Fig. 2.10, the situation is different. The southern area is humid while the northern side is dry, and both of them could be described as lower quality in objective terms compared with an ideal 50% relative humidity. From this perspective, both S-N and N-S groups migrated from one poor-quality to another poor-quality humidity environment, and there was no adaptation—both groups gave a higher score to the region from where they originated.

Next, we want to examine the dynamic patterns of how humidity adaptation is influenced by the time since the migration of the N-S and S-N groups. Figure 2.10 shows different groups' demand factor from the view of humidity perception. As shown in the figure, the demand factors of N-N and S-S groups were slightly lower than those of N-S and S-N. Immediately following migration, the DF of N-S and S-N both increased, showing their initial difficulty in adapting to the newer humidity, in either direction. As the time after migration increased, both N-S and S-N's demand factors decreased with time gradually, eventually getting close to, but still slightly higher, than people who have always lived in the same region. It appears that adaptation to new indoor humidity conditions is more difficult than for changes in indoor temperatures.

2.3.4 *Overall Acceptability*

The radar plots in Fig. 2.11 compares the comfort rate of different groups to the temperature, humidity, and overall acceptance rate on thermal environment. It shows that the overall acceptance rate on thermal environment is similar to their comfort rate of temperature because these two radars overlapped with each other. The comfort rate

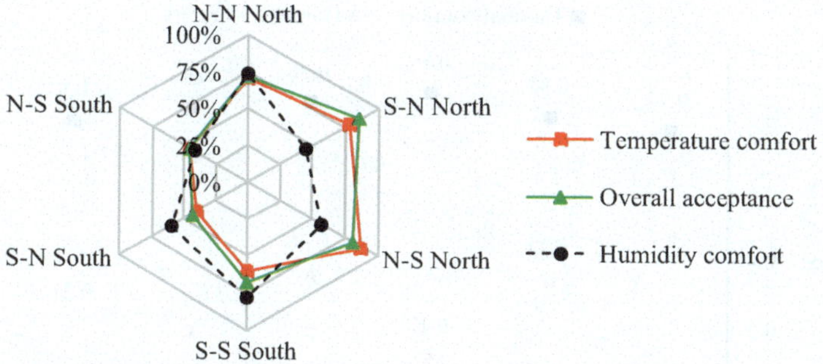

Fig. 2.11 Comfort rate on temperature, humidity, and overall acceptability

radar for humidity is quite different from the other two. It can be qualitatively judged that the overall acceptability is more closely related to temperature conditions.

To better understand the extent to which temperature and humidity comfort influence people's overall acceptance, we used correlation analysis. First, to make the results more intelligible, the votes from 'just uncomfortable' to 'uncomfortable' votes were recoded as '***uncomfortable side***' with a value of '**−1**'; votes from 'just comfortable' to 'comfortable' were grouped into '***comfortable side***' with a value of '**+1**'; and the 'unclear' votes were assigned a neutral value of '**0**'. At the same time, the overall acceptance votes were also simplified in a similar way. The votes from 'very acceptance' to 'just acceptance' were collapsed into a single category, '***acceptable side***' with a value of '**+1**'; votes from 'very unacceptance' to 'just unacceptance' were recoded into '***unacceptable side***' with a value of '**−1**'; and the 'neutral' votes were assigned a value of '**0**'. Then, correlations between overall acceptance rate and temperature and humidity comfort were analyzed, using the correlation coefficient r value to reflect the degrees of correlation between variables. Table 2.8 lists the typical r values and the corresponding degrees of correlation.

The results are shown in Fig. 2.12, which suggests that there is a high correlation between overall acceptance and thermal comfort, where the correlation coefficients are generally above 0.8. Meanwhile, humidity comfort and overall acceptance showed a low correlation, with r values less than 0.5. This means that temperature had a higher influence on overall acceptability in the cases of space heating issues in China.

According to human body heat balance equation, as shown in Eq. 1.1 in Chap. 1, the influence of temperature and humidity on human body heat dissipation can be quantitatively analyzed. The result is shown in Fig. 2.13. In cold condition, tem-

Table 2.8 Correlation coefficient and degree of linear correlation

$\|r\|$ value	1	0.8–1	0.3–0.8	0–0.3	0
Degrees of correlation	Completely correlated	Highly correlated	Moderate correlated	Low correlated	Uncorrelated

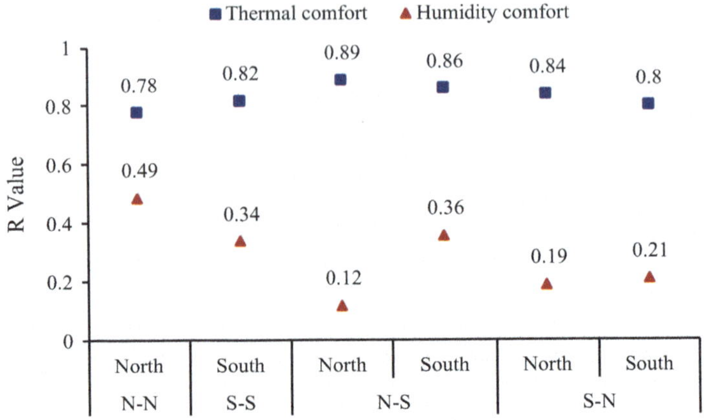

Fig. 2.12 Correlation between thermal/humidity comfort and overall satisfaction

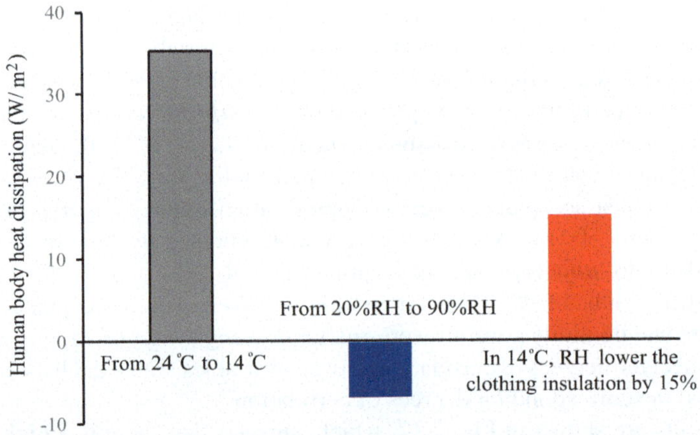

Fig. 2.13 Effect of temperature and humidity change in human body heat dissipation

perature decrease greatly increased human body heat dissipation. When the ambient temperature decreases from 24 to 14 °C, human body heat dissipation increases by 35.2 w/m^2. The increase of relative humidity will reduce the heat dissipation of the human body. When the relative humidity increased from 20 to 90%, human body heat dissipation will be reduced by 6.8 w/m^2. It's worth noting that too high relative humidity will increase the moisture content in clothing, thus reducing the clothing insulation. Assuming the ambient relative humidity increased from 20 to 90% at 14 °C, it will lead to a 15% decrease in the clothing insulation, and this will lead to 15.3 w/m^2 in human body heat dissipation.

2.4 Discussion

2.4.1 The Asymmetry of Thermal Adaptation

Thermal comfort adaptation theory encompasses the idea that people's perception of thermal comfort is influenced by what kinds of thermal environments they have been exposed to. The comparison between group N-N and group S-S in this current study offers some insights into this hypothesis. Group N-N lived in what would be considered a 'higher-grade' winter indoor climate (i.e., it was heated), and the majority of them (57%) expressed neutral thermal sensations, with the remainder roughly divided between warm and cold sensations. This was in stark contrast to group S-S, who lived in southern China without district heating, and 74% expressed cold thermal sensations, and only 23% neutral. In spite of these differences in thermal sensations, however, there were much smaller differences in comfort perceptions of these two groups (71% comfortable for N-N and 60% for S-S). One explanation for this phenomenon is that the S-S group was thermally adapted to these cooler sensations [28, 29].

But, this was not the whole story. Another possible explanation might be the elevated thermal expectations of the N-N group. Table 2.6 shows that, compared with group S-S, the N-N group had a higher uncomfortable rate for almost every thermal sensation category. This suggests that northerners had become increasingly "fussy" about their thermal environment, compared to their southern counterparts who lived in environments with much colder temperatures.

If people with higher thermal expectation and their counterparts with lower expectation, can both be thermally satisfied, how should this convergence be achieved? Should one improve the quality of the indoor thermal environment which the lower expectation group is exposed to, or should one try to lower the expectation of those who have been 'spoiled' by thermal excellence? In other words, which trajectory is easier or might simultaneously lead to reductions in energy use? The comparison between the migrating N-S and S-N groups in the current study suggest that it is much easier to raise one's expectation by providing better thermal environment than it is to persuade those who expected indoor thermal excellence to reduce their expectations and adapt to lower-grade indoor thermal environments. Specifically, both N-S and S-N groups considered winter indoor climate in northern China were more neutral and comfortable than those in southern areas. Apparently, it was easier for the S-N group migrating from environments of lesser quality to indoor climatic excellence, than for the N-S group who migrated in the opposite direction. However, this can also have enormous consequences for increased energy consumption and the risks of climate change.

In the behavioral economy research field, there is a notion called "endowment effect", which describes the tendency for people who own a particular good to value it more than people who do not [30]. Usually, the endowment effect can be attributed to people's loss aversion, which assumes the psychological impact of a loss is greater than an equivalent gain [31]. Under the context of this current study, if we consider the indoor thermal environment as a type of good, it can be seen that building occupants' thermal comfort expectations of the indoor thermal environment can be subject to this loss aversion assumption as well. Migrating from a higher indoor thermal environmental quality to places of lesser indoor thermal environment quality (i.e., the N-S group), the thermal comfort effects are overwhelmingly negative. On the flip-side, migrating from a lesser indoor thermal environmental quality to places of higher indoor thermal environment quality (i.e., the S-N group), the thermal comfort benefits are overwhelmingly positive. Even for the humidity perception, both N-S and S-N groups expressed significant negative feedback on humidity comfort perception when they migrated from one kind of lower-grade indoor humidity environmental to another type of lower-grade indoor humidity environment.

2.4.2 *Practical Implication*

The current study sheds some light on the fact that people's understanding of thermal comfort is, indeed, malleable and is ultimately a subjective state of mind that depends on not only on the physical thermal environments but also social and cultural expectations. People in higher quality thermal environments have been shown to adapt to thermal homogeneity if that is what they expect to have (as for the N-N group). Alternatively, people can also adapt to a more variable indoor climate if that is what they are exposed to repeatedly (as for the S-S group). This phenomenon suggests that building occupants' thermal comfort perception is closely related to their thermal experience, and we believe that comfort expectation plays a factor.

The malleable nature of thermal comfort expectation suggests one might consider alternative policy approaches besides always aiming to build or retrofit buildings with tightly controlled indoor thermal environments. But this approach requires us to rethink how we define the way to provide thermal comfort for building occupants since our current standards and design practices have more typically denied opportunities for occupants to control their own environments to cope with their individual preferences. For example, the current comfort standards like ISO7730 require indoor thermal environments to be tightly controlled around a theoretical optimum by minimizing temporal and spatial fluctuation ranges. Such stringent requirements of indoor thermal environments will undoubtedly shift occupants' perceptions of thermal comfort in the long-term.

Moving forward, we need to advocate for greater flexibility in comfort strategies [32, 33]. To some extent, the adaptive comfort model has made efforts to encourage occupants' adaptive behaviors in the context of naturally ventilated buildings, but there are so many more potential applications. It has been discovered that greater degrees of personal environmental control have positive impacts through both psychological and behavioral adaptive approaches, which could enhance occupants' satisfaction with their indoor thermal environment. If the building's HVAC systems could be running in a "part-time and part-space" mode depending on the occupants' individual needs, instead of the "whole-time and whole-space" mode prevalent in many buildings today, energy use could also be reduced.

2.4.3 Limitations and Future Challenges

The current study is based on a general online survey, and no physical thermal parameters were measured. The issues of adaptation and migration could be further explored through well-designed field studies which combine physical measurements are "right now" surveys (i.e., rather than general thermal impressions). Especially for the demand factor (Eq. 2.2), it can be defined as in sum of uncomfortable percentage throughout a certain temperature range instead of 7-points thermal sensation votes. In this way, more solid conclusions can be drawn with more detailed observations of actual thermal conditions, occupants' activity levels and clothing insulation, personal control activities, etc.

2.5 Conclusion

This study investigated the relationships between occupants' indoor thermal experience and their thermal comfort perception. The following findings and suggestions emerged: (1) People's understandings of thermal comfort are mutually dependent on their indoor thermal experiences. Long-term comfortable thermal experiences can raise occupants' thermal expectation, while lower-grade non-neutral thermal environment can bring about thermal adaptation. (2) Building thermal comfort expectations exhibit asymmetric dynamics. It is easier and quicker for occupants to get accustomed to the thermally neutral lifestyle than it is for 'spoiled' occupants to lower their expectations and adapt to non-neutral indoor climate. (3) It is necessary to review the indoor environmental quality assessment in current standards—their tight requirements may not necessarily produce thermal comfort satisfaction. More flexible approaches and new comfort strategies should be encouraged.

References

1. Banham R (1969) The architecture of the well-tempered environment. The Architectural Press, London
2. Kempton W, Feuermann D, McGarity A (1992) I always turn it on supper: user decisions about when and how to operate air conditioners. Energy Build 18:177–191
3. Shove E, Chappells H (2003) Future comforts: re-conditioning urban environments. Department of Sociology, Lancaster University, UK, Lancaster, UK

4. Healy S (2008) Air-conditioning and the 'homogenization' of people and built environments. Build Res Inf 36(4):312–322
5. Wilhite H (2009) The conditioning of comfort. Build Res Inf 37(1):84–88
6. Mavrogianni A, Johnson F, Ucci M et al (2013) Historic variations in winter indoor domestic temperatures and potential implications for body weight gain. Indoor Built Environ 22(2):360–375
7. Meyer W (2002) Why indoor climates change: a case study. Clim Change 55(3):395–407
8. Hasegawa K, Yoshino H, Ishikawa Y, Matsumoto S (2005) Regional characteristics transition of winter thermal performance and occupants' behavior of detached houses in Tohoku city area for 20 years. J Environ Eng Archit Inst Jpn 593:33–40 (In Japanese)
9. Yoshino H, Yoshino Y, Zhang Q et al (2006) Indoor thermal environment and energy saving for urban residential buildings in China. Energy Build 38(11):1308–1319
10. Cao B, Luo M, Zhou X, Li M et al (2016) Too cold or too warm? A winter thermal comfort study in different climate zones in China. Energy Build 133:469–477
11. Arens E, Humphreys M, de Dear R, Zhang H (2010) Are 'class A' temperature requirements realistic or desirable? Build Environ 45:4–10
12. McIntyre DA (1981) Design requirements for a comfortable environment. Stud Environ Sci 10:195–220
13. Fountain M, Brager G, de Dear R (1996) Expectations of indoor climate control. Energy Build 24:179–182
14. Brager GS, de Dear RJ (1998) Thermal adaptation in the built environment: a literature review. Energy Build 27(1):83–96
15. Fanger PO, Toftum J (2002) Extension of the PMV model to non-air-conditioned buildings in warm climates. Energy Build 34(6):533–536
16. Nicol J, Humphreys M (2002) Adaptive thermal comfort and sustainable thermal standards for buildings. Energy Build 34(6):563–572
17. Rajasekar E, Ramachandraiah A (2010) Adaptive comfort and thermal expectations—a subjective evaluation in hot humid climate. In: Proceedings of conference: adapting to change: new thinking on comfort cumberland lodge, Windsor, UK
18. Andamon MM, Williamson TJ, Soebarto VI (2006) Perceptions and expectations of thermal comfort in the Philippines. In: Proceedings of Windsor 2006 conference: comfort and energy use in buildings—getting them right. Windsor, UK
19. Amin R, Teli D, James P, Bourikas L (2016) The influence of a student's 'home' climate on room temperature and indoor environmental controls use in modern halls of residence. Energy Build 119:331–339
20. Wang Z, Ji Y, Ren J (2017) Thermal adaptation in overheated residential buildings in severe cold area in China. Energy Build 146:322–332
21. Luo M, de Dear R, Ji W et al (2016) The dynamics of thermal comfort expectations: problem, challenge and implication. Build Environ 95:322–329
22. Ning H, Wang Z, Ji Y (2016) Thermal history and adaptation: does a long-term indoor thermal exposure impact human thermal adaptability? Appl Energy 183:22–30
23. Luo M, Cao B, Zhou X (2014) Can personal control influence human thermal comfort? A field study in residential buildings in China in winter. Energy Build 72:411–418
24. Li B, Tan M, Liu H et al (2010) Occupant's perception and preference of thermal environment in free-running buildings in China. Indoor Built Environ 19(4):405–412
25. Han J, Yang W, Zhou J et al (2009) A comparative analysis of urban and rural residential thermal comfort under natural ventilation environment. Energy Build 41(2):139–145
26. Sawilowsky SS (2009) New effect size rules of thumb. J Mod Appl Stat Methods 8(2):597–599
27. Luo M, Cao B, Ouyang Q, Zhu Y (2017) Indoor human thermal adaptation: dynamic processes and weighting factors. Indoor Air 27(2):273–281
28. Liu W, Zheng Y, Deng Q, Yang L (2012) Human thermal adaptive behavior in naturally ventilated offices for different outdoor air temperatures: a case study in Changsha China. Build Environ 50:76–90

29. Li B, Yu W, Liu M, Li N (2011) Climatic strategies of indoor thermal environment for residential buildings in Yangtze River Region, China. Indoor Built Environ 20(1):101–111
30. Morewedge C, Giblin C (2015) Explanations of the endowment effect: an integrative review. Trends Cogn Sci 19(6):339–348
31. Kahneman D, Tversky A (1979) Prospect theory: an analysis of decision under risk. Econometrica 47:263–292
32. Ghahramani A, Tang C, Becerik-Gerber B (2015) An online learning approach for quantifying personalized thermal comfort via adaptive stochastic modeling. Build Environ 92:86–96
33. Ghahramani A, Castro G, Becerik-Gerber B, Yu X (2016) Infrared thermography of human face for monitoring thermoregulation performance and estimating personal thermal comfort. Build Environ 109:1–11

Chapter 3
The Timescale of Thermal Comfort Adaptation in Heated and Unheated Buildings

3.1 Background

3.1.1 Dynamic Characteristics of Thermal Comfort Adaptation

Despite the progress that has been made in adaptive thermal comfort research, there remain some unproven premises. One typical uncertainty is the adapting direction: 'if building occupants can adapt to different indoor environments, being more forgiving in naturally ventilated buildings and fussier about air-conditioned environments, what is the response of those who move between different building types? Is it possible for those who are accustomed to one building type to adapt to another building type?' These questions go to the very core of adaptive theory—the mutually dependent relationship between indoor climate experience and thermal adaptation. Understanding the answers to these questions can help nail down the ambiguous situation of thermal comfort evaluation of mixed-mode buildings. Given that mixed-mode buildings usually have mechanical air conditioning that coexists with natural passive cooling, it is very complicated to determine its interior thermal comfort. Although some field studies [1, 2] indicated that mixed-mode building occupants' thermal comfort response is more in line with naturally ventilated buildings, to date, there is no consensus on the thermal comfort approach that should be applied to mixed-mode buildings. More empirical evidence is needed to determine whether they should be considered naturally ventilated buildings using the adaptive model or air-conditioned buildings using the heat balance model [3]. Taking a broader perspective, the above questions can be expressed in higher-level versions: 'if people became accustomed to the 'ideally comfortable' indoor thermal conditions created by mechanical HVAC systems, can they adapt to underconditioned indoor environments easily and quickly? [4, 5]

M. Luo, *The Dynamics and Mechanism of Human Thermal Adaptation in Building Environment*, Springer Theses,
https://doi.org/10.1007/978-981-15-1165-3_3

When adapting to different indoor climates, do they show the same trajectory?' All these questions point toward another uncertainty—the dynamic characteristics of thermal adaptation. The significance of these seemingly simple questions cannot be overstated, as global demographic superpowers such as Brazil, India, and China, are embarking upon tremendous construction boom in the history of humankind. The type of building environments these countries will construct will largely affect future global energy consumption and greenhouse gas emissions.

With limited direct empirical evidence, the relationship between indoor climate experience and people's thermal adaptation together with the dynamic characteristics of thermal adaptation have been anecdotal. This may be due in large part to its personal nature and complexity. Similar to the evolution and layers-of-the-onion metaphors, thermal adaptation involves many confounding factors such as physical, physiological, psychological, and even cultural and economic differences. No consensus has been reached on whether the indoor or outdoor climate shapes a person's thermal adaptation. By analyzing a large amount of raw data compiled from the ASHRAE thermal comfort database I [6], de Dear and Brager [7] interpreted thermal adaptation in naturally ventilated buildings as behaviorally and psychologically adapting to the varied indoor climates driven by the external weather and the seasonal cycles. Since then, many adaptive models and equations have been developed, taking outdoor climate conditions as the primarily driving force [8–10]. On the other hand, some studies found that indoor climate experience also plays an important role in shaping occupants' thermal adaptation. For example, Yu et al. [11] compared two groups' (one group was acclimated to air-conditioned environments while the other group was acclimated to naturally ventilated environments) phycological responses when adapting to warm conditions. It was found that the natural ventilation (NV) group had a stronger capability for physiological thermoregulation than air conditioning (AC) group. van der Lans et al. [12] found that a 10-day cold acclimation protocol can increase brown adipose tissue activity in parallel with an increase in non-shivering thermogenesis in humans. The subjects expressed less perception of cold, fewer complaints of discomfort, and fewer shivering reports when exposed to cold conditions after 10 days of consecutive cold acclimation (15–16 °C). More empirical evidence needed to determine how thermal experience affects building occupants' thermal adaptation.

Even fewer studies have examined the dynamic characteristics of thermal adaptation. Generally, studies focused on this topic can be sorted into two groups depending on different adaptation time scales. The short-term group investigated thermal adaptation on a scale of days, weeks, or months. For example, Candido et al. [13] studied the influence of short-term (successive days) of hot-humid acclimation on subjects' perception of the indoor thermal environment. They found that even though subjects were constantly exposed to air-conditioned indoor environments, the artificially induced heating acclimation improved their thermal and air movement assessment. Liu et al. [14] conducted a short-term (8-day) tracking field study of 7 Chinese migrants who traveled from Xi'an (with a cold, dry climate) to Qionghai (with a hot, humid climate). Interestingly, their results showed that those migrants adapted to the

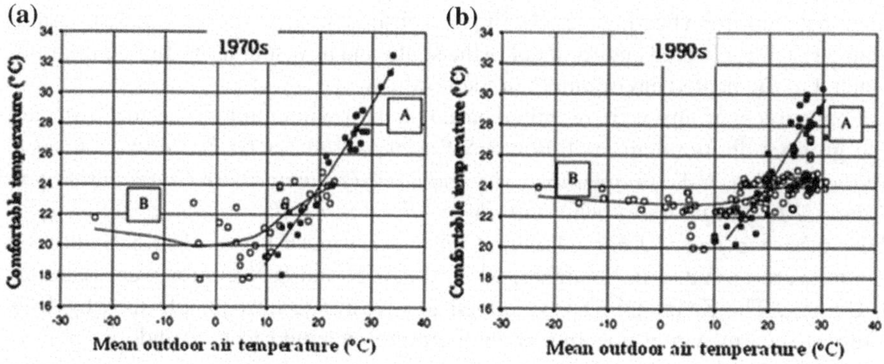

Fig. 3.1 Comfort temperature as a function of outdoor temperature, **a** for free-running buildings and **b** for conditioned buildings [16]

local climate in approximately three days. Meanwhile, the long-term group analyzed thermal adaptation on a scale of seasons, years, or even decades. One such example is the study by Nicol and Humphreys [15], in which they found the comfortable indoor temperatures in heated or cooled office buildings in the 1990s (shown in Fig. 3.1b) were much more tightly clustered than those observed in 1970s (shown in Fig. 3.1a). These efforts helped better understand the dynamics of thermal adaptation, but they are either a laboratory experiment or a web survey. Well-designed thermal comfort field studies in real buildings have yet to be conducted to validate their findings.

3.1.2 Objectives of This Chapter

This chapter introduces two comparative long-term thermal comfort field studies, attempting to answer the following questions: (1) How do people's indoor thermal experience influence their thermal adaptation? (2) How can we describe the dynamic characteristics of thermal adaptation?

3.2 Method

3.2.1 Field Study Design

The generalization of wintertime indoor temperature difference between northern and southern China (as described in Chap. 2) offered a natural opportunity for comparative research on the dynamics of thermal comfort expectation. We selected two representative cities: one located in the southern unheated area (Shanghai) and the other in the northern heating area (Beijing) to observe how northern Chinese migrants adapted to the unheated environment in the south and how southern Chinese migrants adapted to the heated environment in the north.

Four subject groups were recruited based on their winter indoor thermal histories. The group definitions are as follows: *S-S group—those who had always lived in southern China; N-S group—those who had moved from northern China to southern China; N-N group—those who had always lived in northern China; S-N group—those who had moved from southern China to northern China.* The S-S and N-S groups were recruited in Shanghai, while the N-N and S-N groups were recruited in Beijing. The Shanghai surveys aimed to investigate how people moving from neutral-to-warm wintertime indoor temperatures adapted to cold indoor environments. The Beijing surveys were designed to determine how people from unheated regions adapted to space-heated building environments.

To study the dynamic process of building occupants' thermal adaptation and their transformation in comfort expectation, the ideal research method is a long-term tracking survey that can continuously document respondent' thermal comfort histories over a long period of time (perhaps years). In this way, changes during different adapting phases can be recorded. However, long-term tracking requires years and a large amount of labor cost, which makes it almost impossible. In the current study, we chose an alternate approach—subgrouping N-S and S-N subjects according to how much time had elapsed since their migration. For example, we subgrouped S-N subjects into S-N-1, S-N-2, and S-N-3 groups. S-N-3 group consists of subjects who grew up in southern China with cold wintertime indoor temperatures but had lived in Beijing for 3 years. The subjects in the S-N-2 group experienced a similar indoor thermal history as the S-N-3 group but lived in Beijing for 2 years. The S-N-1 group contains those subjects who moved from southern China to Beijing and had been in Beijing for less than 1 year.

To better control the groups with different migrating times, all subjects were healthy college-aged students with similar proportions of males and females. They all had light-to-none caffeine, alcohol, smoking habits—less than 2 cups of coffee or 2 cigarettes per day, and normal exercise intensity 2–4 times per week. The final valid responses for each group and more detailed information about the participants are listed in Table 3.1.

To take advantage of the wintertime indoor temperature difference in southern and northern China, we conducted most of the comparative field studies during winter to ensure the indoor thermal generalizations. We also tried to include wider indoor temperature ranges as much as possible. The Shanghai studies were conducted from September 2012 to May 2013, with average daily outdoor temperature ranging from −2.5 to 30.2 °C. The Beijing surveys were conducted from October 2015 to April 2016, with average daily outdoor temperature ranging from −13.5 to 28.8 °C.

Table 3.1 Subject information

Location	Group		Indoor thermal experience	Sample size		Age (years)	Valid response
				Male	Female		
Shanghai	S-S		Grew up in southern China without space heating, and has moved to Shanghai	26	16	20.5	1033
	N-S	N-S-1	Grew up in northern China with space heating, and moved to Shanghai for <1 year	13	9	21.5	201
		N-S-2	Grew up in northern China with space heating, and moved to Shanghai for 2 years	14	6	20.2	425
		N-S-3	Grew up in areas with space heating, and moved to Shanghai for 3 years	16	5	21.3	441
Beijing	N-N		Grew up in northern China with space heating, and moved to Beijing	14	10	20.9	223
	S-N	S-N-1	Grew up in southern China without space heating, and moved to Beijing for <1 year	5	7	19.8	134
		S-N-2	Grew up in southern China without space heating, and moved to Beijing for 2 years	7	6	20.4	126
		S-N-3	Grew up in southern China without space heating, and moved to Beijing for 3 years	4	7	21.8	151

3.2.2 Physical Parameters Measurement

Figure 3.2 shows the field scenes of the surveyed buildings. The Shanghai case study buildings are teaching buildings constructed in 2006 which are installed with air conditioning systems. Generally, the buildings run in natural ventilation mode, but space heating or air conditioning are turned on during extremely cold or hot

a. Case buildings in Shanghai b. Case buildings in Beijing

Fig. 3.2 Investigation of building appearance and site conditions

days. The buildings surveyed in Beijing are student dormitories built in 2003 which are equipped with district heating systems. Typically, the heating season is from November 1 to March 15 of the following year.

Table 3.2 lists the survey instruments for measuring physical environment parameters. In the Shanghai survey, for example, parameters such as indoor air dry-bulb temperature, relative humidity, and black globe temperature were continuously monitored at 5-min intervals. As the indoor air velocity was relatively small, usually below the sensitivity threshold of 0.2 m/s, it was measured intermittently instead of continuously. To capture the representative thermal parameters in the surveyed rooms, we fixed the measuring points at 0.6–1.1 m in height representing the main sitting body activity areas, and at multiple locations within 2 m of the subjects.

Table 3.2 Environmental monitoring instruments

Equipment	Measurement parameters	Scale	Precision
Temperature and humidity recorder (WSZY-1)	ta	5–35 °C	±0.25 °C
	RH	20–80%	±3%
Black globe temperature recorder (HWZY-1)	tg	5–35 °C	±0.25 °C
Anemometer (FB-1)	v	0.05–5 m/s	±5%
Thermal comfort recorder (SSDZY-1)	ta	5–35 °C	±0.25 °C
	RH	0–100%	±3%
	tg	5–35 °C	±0.25 °C
	v	0.05–5 m/s	±5%

3.2.3 Subjective Questionnaire

In addition to measuring the physical thermal environmental parameters, a simultaneous subjective survey was conducted, which included the following three aspects.

Subject background information. In the survey preparation stage, basic subject information such as age, gender, hometown, the use of space heating facility in winter, time since migrating, etc. was collected. Next, we purposely selected subjects who met the criteria listed in Table 3.1.

Clothing insulation and activities. This part of the questionnaire was designed to collect subjects' clothing insulation and activity levels. The clothing insulation information calculated in accordance with ASHRAE Standard 55-2013 and was recorded. Certain clothes with Chinese characteristics were also taken into consideration. At the same time, we asked subjects' activity statuses using the questionnaire and then obtained the activity's corresponding metabolic rate according to the suggested values in ISO 8996 [17]. For example, a common seating posture with mild mental work is considered to have a metabolic rate of 1.1 met (1 met = 58.2 W/m^2). To determine whether the metabolic rate was stable, subject activity levels during the past 30 min were questioned.

Subjective comfort evaluation. To record the subjects' evaluation of the indoor thermal environment, parameters such as thermal sensation vote (TSV), thermal comfort vote (TCV), thermal acceptance, and thermal preference were determined by questionnaire. The TSV adopted a seven-point scale ranging from cold (−3) to hot (+3), with neutral (0) in the middle. The thermal comfort and acceptability questions also followed the same pattern. Thermal preferences were assessed using a three-point scale: 'warmer', 'no change', and 'cooler'. These question scales are consistent with the global thermal comfort database [18].

3.2.4 Data Processing

Data Standardization. To compare the results from different surveys, certain modifications were made to reduce the number of confounding factors. First, we standardized the relative humidity, air velocity, clothing insulation, and metabolic rate parameters, but used the original air temperature and mean radiant temperature measurements. The standardized parameter descriptions are shown in Table 3.3. Then, a correction was made for each respondent in four steps: (a) a PMV value was calculated according to its actual measured parameters (shown in Eq. 3.1); (b) a standardized PMV, expressed as PMV', was calculated using standardized parameters (shown in Eq. 3.2); (c) the ΔPMV was calculated by subtracting the PMV' from the actual PMV (shown in Eq. 3.3); (d) the modified TSV' was obtained by subtracting ΔPMV from the actual TSV (shown in Eq. 3.4).

$$\mathrm{PMV} = f\left(\mathrm{t_{op}}, \mathrm{RH}, \mathrm{v}, \mathrm{I}, \mathrm{M}\right) \tag{3.1}$$

Table 3.3 Standardized parameters

Parameters	Description
T_a (°C)	Actual measured data
T_r (°C)	Actual measured data
RH (%)	Averaged data during a survey period
v (m/s)	Averaged data during a survey period
I_{cl} (clo)	Averaged value in each binned temperature
M (met)	1.1 (Typical teaching building activity—sedentary learning)

$$\mathrm{PMV}' = f\left(\mathrm{t_{op}}, \overline{\mathrm{RH}}, \bar{\mathrm{v}}, \bar{\mathrm{I}}, \overline{\mathrm{M}}\right) \quad (3.2)$$

$$\Delta\mathrm{PMV} = \mathrm{PMV} - \mathrm{PMV}' \quad (3.3)$$

$$\mathrm{TSV}' = \mathrm{TSV} - \Delta\mathrm{PMV} \quad (3.4)$$

Statistical analysis. For better comparison among groups, descriptive statistics such as the mean and the standard deviation were calculated. When comparing the results from different groups, two sample *T*-tests were carried out to validate the statistical significance. The outcomes were interpreted as follows: $p \leq 0.001$ or "***" means highly significant, $0.001 < p \leq 0.01$ or "**" means significant, $0.01 < p \leq 0.05$ or "*" means weakly significant, and $p > 0.05$ means not statistically significant. To determine whether thermal sensation varies significantly between different subgroups, two-way ANOVA tests with migration time and binned temperature as the main factors were performed. If there were significant main effects or interactions, Tukey's HSD tests were applied to identify which interaction caused the difference. Usually, $p \leq 0.05$ was considered the statistical significance criterion.

Demand factor calculation. To demonstrate thermal adaptation trajectories, we adopted the 'demand factor (DF)' index proposed in our previous chapter to compare different groups' adaptation processes. The basic idea of the DF index is: if a group of people has relatively high requirements or expectations of the indoor thermal environment and lower tolerances to uncomfortable conditions that deviate from thermal neutrality, these factors will collectively lead to a larger DF value. Alternatively, people who have relatively less-stringent thermal environment requirements and higher tolerances for uncomfortable thermal conditions will have a smaller DF. According to this logic, DF was originally defined as shown in Eq. 3.5, where 'i' indicates $\mathrm{group_i}$, '$\sum_{\mathrm{cold}}^{\mathrm{hot}}$ percentage of $\mathrm{uncomfortable_i}$' indicates the sum of $\mathrm{group_i}$'s uncomfortable percentages across the range of cold-to-hot thermal sensation votes, and '$\sum_{\mathrm{cold}}^{\mathrm{hot}}$ percentage of $\mathrm{uncomfortable_{baseline}}$' indicates the sum of $\mathrm{group_{baseline}}$'s uncomfortable percentages across the range of thermal sensation votes.

$$\text{Demand factor}_i = \frac{\sum_{cold}^{hot} \text{percentage of uncomfortable}_i}{\sum_{cold}^{hot} \text{percentage of uncomfortable}_{baseline}} \quad (3.5)$$

However, the previous chapter was based on an online general thermal impression survey; no physical thermal parameters were measured. The current study is a well-designed "right-here-right-now" field survey combined with physical measurements. For this reason, the DF index can also be defined as the sum of the uncomfortable percentages in a certain temperature range rather than as a 7-point scale of thermal sensation votes. The revised DF definition is as shown in Eq. 3.6, where '$\sum_{14}^{28}$ percentage of uncomfortable$_i$' represents the sum of group$_i$'s uncomfortable percentages across the range of temperature bins from 14 to 28 °C, and '$\sum_{14}^{28}$ percentage of uncomfortable$_{baseline}$' means the sum of group$_{baseline}$'s uncomfortable percentages across the range of 14–28 °C.

$$\text{Demand factor}_i = \frac{\sum_{14}^{28} \text{percentage of ncomfortable}_i}{\sum_{14}^{28} \text{percentage of uncomfortable}_{baseline}} \quad (3.6)$$

For comparison, we used both Eqs. 3.5 and 3.6 to calculate the DF index. When calculating the DF index, the selection of the baseline (or reference) group is flexible. It can be a real group of people or a hypothetical group determined using the PMV-PPD model. In this study, we selected the average of all the respondents as the baseline group, which includes the S-S, N-N, S-N and N-S subgroups. When DF is > 1, group$_i$ had higher levels of discomfort, or a higher demand or expectations compared to the baseline group, and the opposite is true when DF < 1.

Tools used. The raw data were prepared and cleaned using Microsoft Excel 2016. Statistical calculations and significance analyses were performed in R (Version 3.5). Some of the figures were made in OriginPro (Version 2018, OriginLab, Northampton, MA, USA).

3.3 Results

3.3.1 Measured Indoor Thermal Parameter

Figure 3.3 shows the measured indoor and outdoor temperatures during the survey period. Figure 3.3a shows the measured temperatures in naturally ventilated buildings in Shanghai with no district heating capacity. The indoor temperatures were closely correlated with outdoor temperatures, indicating that the surveyed buildings were in a good state of natural ventilation. Additionally, due to the long-time span of the survey, the monitored indoor and outdoor temperatures were widely distributed. The indoor temperatures ranged from 10.1 to 29.1 °C, and the corresponding outdoor temperatures ranged from 2.9 to 33.4 °C.

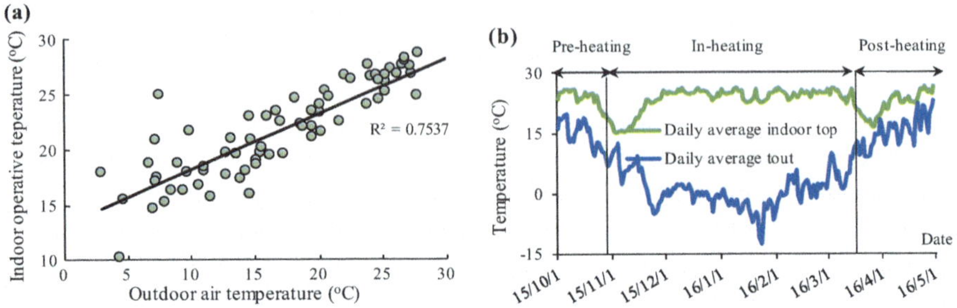

Fig. 3.3 Indoor temperature and outdoor temperature: **a** naturally ventilated buildings in Shanghai; **b** heated buildings in Beijing

Figure 3.3b shows the daily averaged indoor and outdoor temperatures monitored in the Beijing case study buildings. As the investigation time in Beijing began on October 1, 2015, and lasted until April 30 of the following year, it included the entire space heating season from November 1 to March 15 of the following year. As shown in the chart, the temperature variations can be divided into three phases: (1) *Preheating*. The indoor temperature decreased gradually as the outdoor temperature dropped. During this phase, the lowest indoor temperature was 15.1 °C. (2) *In-heating*. The indoor temperature increased rapidly and remained within a stable range of 21–26 °C, even though the outdoor temperature was cold and fluctuated greatly. (3) *Postheating*. When space heating was suspended, the indoor temperature dropped suddenly at the beginning, and then increased again, following a "V" shaped trend.

Table 3.4 lists the statistics of each subgroup's measured and calculated thermal parameters. Because the PMV value can be considered a combined reflection of thermal environmental parameters such as air temperature and air velocity, and other factors such as clothing insulation and metabolic rate, it was used to preliminarily compare the indoor thermal conditions of each subgroup. The average PMV values were approximately −0.7 for the Shanghai groups and −0.13 for the Beijing groups, indicating that the Beijing buildings had warmer indoor thermal environments because of space heating. Although the measured outdoor temperature in Beijing (−12 to 23 °C,) was much colder than Shanghai's 2.9–33.4 °C, the lowest indoor temperature in Beijing was 14.8 °C higher than the 10.8 °C measured in Shanghai.

Figure 3.4 projects each group's indoor thermal conditions on the two classic thermal comfort zones. By presenting the average values of indoor thermal parameters measured in specific surveys, Fig. 3.4 compares the distribution of thermal conditions experienced by different groups. The widely (but randomly) distributed dots in the figure suggest that the three Beijing groups experienced similar indoor thermal conditions as did the Shanghai groups. Independent *T*-test results also indicate no significant difference ($p < 0.05$) among the Beijing subgroups, as well as among the Shanghai subgroups.

Table 3.4 Statistics of thermal environment indices

Parameters		S-S group			N-S-3 group			N-S-2 group			N-S-1 group		
		Mean	Max.	Min.	Mean	Max.	Min.	Mean	Max.	Min.	Mean	Max.	Min.
Shanghai survey	t_{out}	18.0	33.4	2.9	19.4	33.4	7.0	16.5	27.1	2.9	15.3	27.6	2.9
	ta	22.4	29.1	11.9	23.0	29.1	13.0	20.8	28.2	10.8	20.4	27.4	12.1
	RH	62.1	78.6	37.8	59.3	76.5	37.8	57.5	78.6	42.2	60.0	76.5	45.3
	ν	0.08	0.21	0.02	0.09	0.21	0.05	0.08	0.19	0.02	0.08	0.18	0.03
	t_{op}	22.6	28.8	14.8	22.9	28.8	14.8	21.2	28.3	15.5	20.5	26.9	15.5
	I	0.74	1.18	0.36	0.65	1.14	0.38	0.78	1.13	0.37	0.75	1.07	0.33
	PMV	−0.7	1.3	−1.7	−0.7	1.3	−1.8	−0.5	1.0	−1.6	−0.8	1.1	−2.1
		N-N group			S-N-3 group			S-N-2 group			S-N-1 group		
Beijing survey	t_{out}	6	23	−12	5.9	21.5	−12	5.9	21.5	−7	6.4	23	−11
	ta	23.7	26.5	14.8	23.3	26.2	15.3	23.1	26	15.3	22.9	26.5	14.8
	RH	38.1	64.2	16.2	36.9	64.2	16.4	38.2	63.5	16.2	37.1	62.3	16.4
	ν	0.09	0.27	0.02	0.09	0.21	0.05	0.10	0.25	0.02	0.11	0.22	0.05
	t_{op}	23.3	27.1	14.8	23.9	26.8	14.8	23.2	26.3	15.1	23.4	27.1	15.2
	I	0.74	1.52	0.36	0.71	1.14	0.38	0.73	1.13	0.37	0.83	1.07	0.43
	PMV	−0.2	0.5	−1.2	−0.1	0.6	−1.4	−0.3	0.5	−1.1	−0.1	0.7	−1.3

Note The outdoor temperatures were different for different groups because the survey was conducted on different days

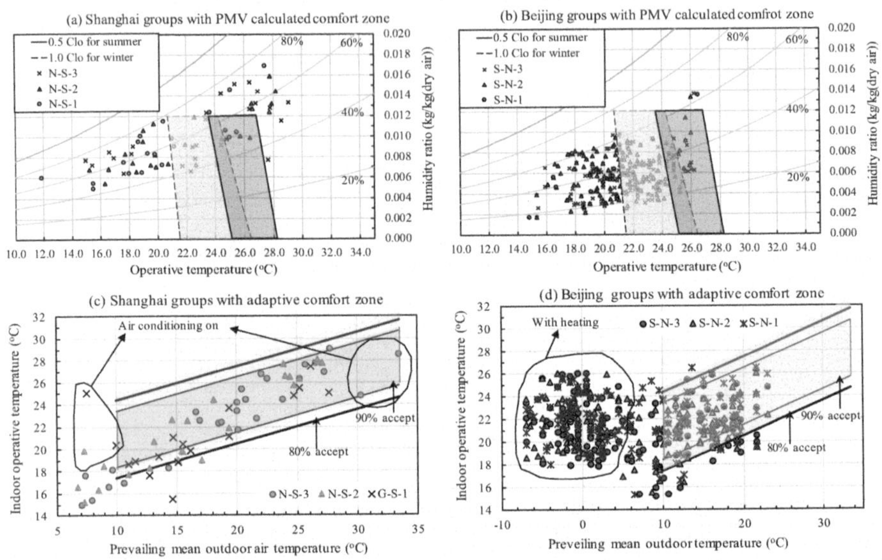

Fig. 3.4 Distributions of indoor thermal conditions. *Note* Because the S-S and N-N groups were surveyed simultaneously with other groups, their thermal environment parameters are summarized by other groups and are not listed separately in the figure. Additionally, most Shanghai surveys were conducted in naturally ventilated conditions, but 5 out of the 64 surveys included air conditioning or space heating. All surveys were included in the data analysis

Comparing the measured data with the adaptive model (Fig. 3.4c, d) or the PMV calculated comfort zone (Fig. 3.4a, b). Figure 3.4 shows the importance of choosing the right comfort model to evaluate an indoor thermal environment. For instance, if the adaptive comfort zone was applied to the Shanghai groups, more than 80% of their thermal conditions can be regarded as comfortable (see Fig. 3.4c). If the PMV calculated comfort zone was applied, fewer than 50% of the measured indoor conditions fall into the recommended comfort zone (see Fig. 3.4a). However, the actual satisfaction rate of those surveys was higher than 90%, which seems to be more in-line with the adaptive comfort approach. Similarly, the Beijing groups had nearly half of their cases out of the PMV calculated comfort zone even though they had space heating (see Fig. 3.4b). This might be because of the preheating and postheating periods, or overheating issues during the space heating period. However, when applying the adaptive comfort zone, the cold outdoor temperatures in Beijing exceed the range of the current adaptive comfort model, which results in more than 60% of the cases falling outside the acceptable zone (see Fig. 3.4d).

3.3.2 *Subjective Comfort Perception*

Figure 3.5a, b show the actual and predicted thermal sensation corresponding to the binned indoor temperatures. Each dot represents the average TSV and PMV in a 1 °C operative temperature interval. The Shanghai groups' TSV regression lines exhibit a crisscross-shaped difference from their PMV regression line (see Fig. 3.5a), which means PMV overestimated the Shanghai groups' cold sensation in cool environment and hot sensation in warm environments. In contrast, the Beijing groups' results showed that the two lines overlapped (see Fig. 3.5b), indicating the PMV can accurately predict the Beijing groups' thermal sensations.

The sensation differences were also reflected in the acceptance rate. Figure 3.5c, d show the Shanghai and Beijing groups' dissatisfaction rates across different temperatures, respectively. The Shanghai groups' (see Fig. 3.5c) actual dissatisfaction curve exhibits an obvious deviation from the predicted curve. Especially in conditions far from thermal neutrality, the predicted dissatisfaction rate is significantly higher than the actual dissatisfaction rate. If 80% satisfaction rate is considered to guarantee occupants' acceptability of an indoor thermal environment, the acceptable

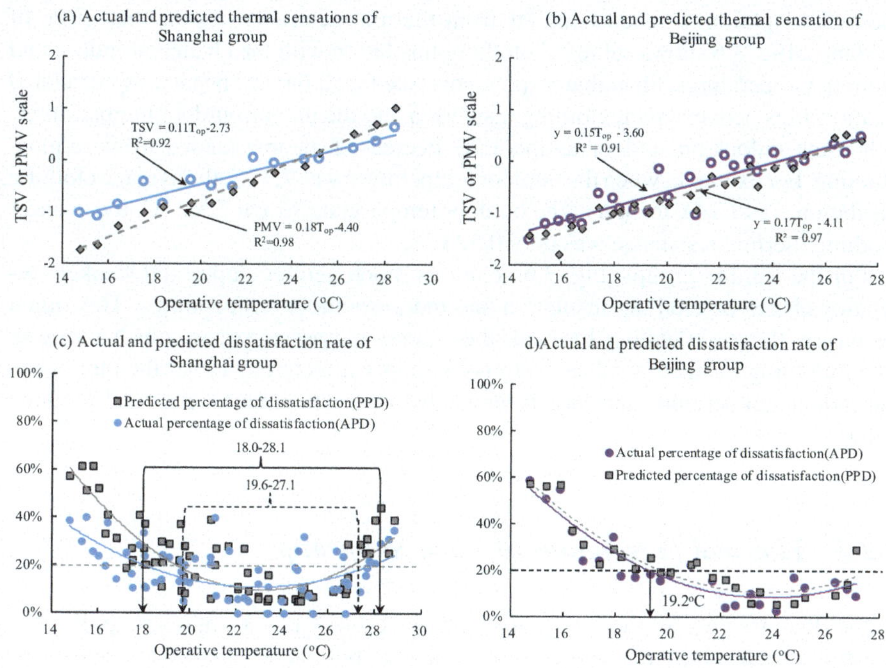

Fig. 3.5 Thermal sensation and thermal acceptance across binned temperature. *Note* The actual dissatisfaction rate was calculated from the acceptance question, while the predicted dissatisfaction rate was based on PMV-PPD calculation

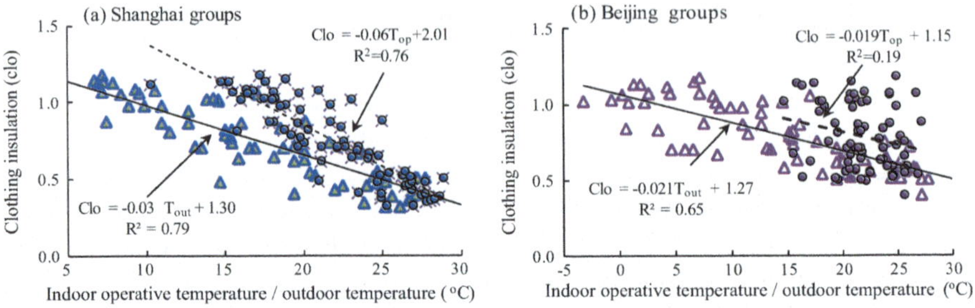

Fig. 3.6 Clothing insulation with indoor and outdoor temperatures. *Note* Each dot represents the average clothing insulation in a specific survey

indoor temperature range is 18.0–28.1 °C from the actual dissatisfaction curve, 2.6 °C wider than the 19.6–27.1 °C range obtained from the predicted curve. On the other hand, the Beijing groups' (see Fig. 3.5d) actual dissatisfaction rate is very close to the predicted rate. The 80% acceptable lower temperature limit is 19.2, 1.2 °C higher than that of the Shanghai group.

The results shown in Fig. 3.5 indicates that the Shanghai groups had a strong thermal adaptation to conditions far from thermal neutrality. As another piece of evidence, Fig. 3.6 shows subjects' clothing insulation with the change of indoor and outdoor temperatures. Shanghai respondents (see Fig. 3.6a) had negatively correlated relationships between their clothing insulation and the indoor/outdoor temperatures. When the indoor or outdoor temperature decreased, people tended to wear more clothing. For example, when the outdoor temperature was 7.3 °C, the average clothing insulation was 1.14 Clo. When the outdoor temperature rose to 25.2 °C, the average clothing thermal resistance was only 0.52 Clo.

For the Beijing groups, Fig. 3.6b exhibits much gentler slopes and weaker correlations between clothing insulation and indoor/outdoor temperatures. This might because of the space heating that made their indoor temperature range much narrower. Their clothing insulation values were not strongly affected by either the outdoor or the indoor temperature, but largely depended on each subject's personal clothing habit.

3.3.3 *Thermal Perceptions of Each Subgroup*

Rather than treating all the subgroups listed in Table 3.1 as a single group, Fig. 3.7 illustrates each subgroup's thermal sensation and thermal acceptance rate across the binned temperatures. Figure 3.7a, c show how the Shanghai groups adapted to the naturally ventilated buildings without winter heating, while Fig. 3.7b, d illustrate the Beijing groups' thermal adaptation after moving to buildings with space heating in winter.

As shown in Fig. 3.7a, when exposed to a mildly cold environment, the N-S-1 group tended to have the lowest thermal sensation vote, followed by the N-S-2 group, the N-S-3 group and lastly the S-S group. Statistical tests (ANOVA plus Tukey's HSD test, as shown in Table 3.5) indicate that when the temperature was below 26 °C, N-S-1 had a significantly lower thermal sensation value than the S-S group; the N-S-2 group also showed a significant difference from S-S group when the indoor temperature was below 20 °C. However, no significant difference was observed between the N-S-3 group and the S-S group. Similar phenomena were found in the thermal acceptance rate, as shown in Fig. 3.7c. It can be inferred that the dynamic process of adapting to an unheated environment is relatively slow, perhaps 2–3 years.

For the Beijing groups, Fig. 3.7b, d show that the S-N-1 group had significantly warmer thermal sensations in mildly cold environments below 22 °C compared to the N-N groups. The S-N-1 group's acceptance rate was slightly higher in mildly cold conditions but lower in the warm environment compared to other groups. No significant difference was found between the N-S-3 or N-S-2 group and the N-N group.

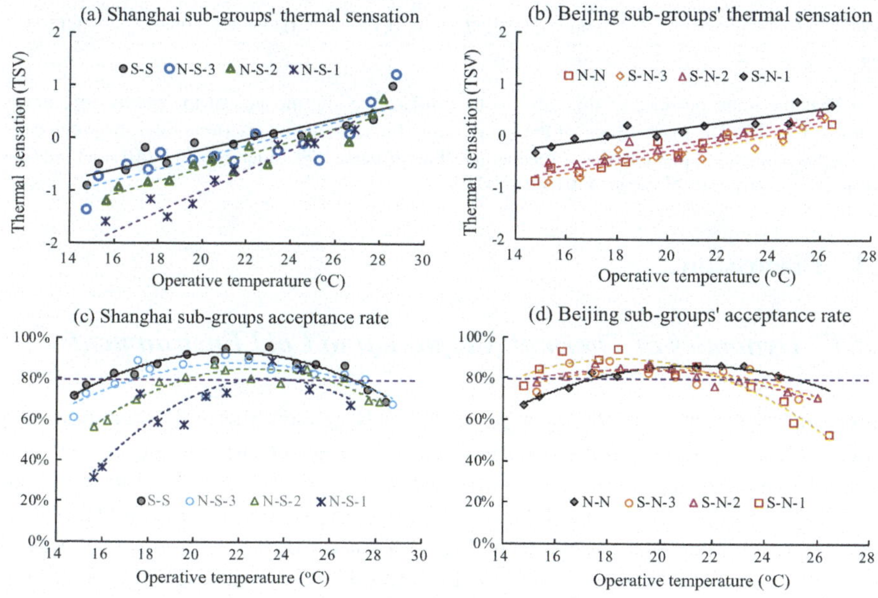

Fig. 3.7 Thermal acceptance rate votes for different subgroups

Table 3.5 Thermal sensation ANOVA with Tukey's HSD test p values

Binned temperature (°C)	S-S			N-N		
	N-S-3	N-S-2	N-S-1	S-N-3	S-N-2	S-N-1
14	0.24	–	–	–	–	–
15	0.59	**	***	0.61	0.62	*
16	0.24	*	***	0.13	0.34	*
17	0.13	**	*	0.28	0.72	*
18	0.61	0.62	***	0.54	0.15	*
19	0.82	*	***	0.89	0.32	0.15
20	0.69	*	***	0.64	0.30	*
21	0.82	0.09	***	0.63	0.27	*
22	0.91	0.56	***	0.99	0.55	*
23	0.93	0.08	***	0.89	0.09	–
24	0.64	0.16	*	0.82	0.86	0.16
25	1.00	0.67	*	0.35	0.96	0.09
26	0.49	0.98	0.06	0.99	0.99	0.29
27	0.99	0.37	0.19	0.99	0.99	0.34
28	0.99	0.57	0.28	–	–	–

Note That the main purpose of this table is to validate the significance of the group differences shown in Fig. 3.7. The main effects of the temperature bins and the interaction effects between each pair of groups are not presented. '–' means insufficient data, '***' means $p \leq 0.001$, '**' means $0.001 < p \leq 0.01$, and '*' means $0.01 < p \leq 0.05$

3.4 Discussion

3.4.1 Dynamics of Thermal Adaptation to Cold Environment

Figure 3.7 compares the N-S group and the S-N group's thermal adaptation to a mildly cold environment. It indicates that the N-S subgroups took longer (2–3 years) to adapt to a naturally ventilated environment during winter in Shanghai, while the S-N subgroups quickly accepted the warm heated environment in Beijing (less than 1 year). To validate this observation, Fig. 3.8 shows different groups' calculated demand factors. The N-N group had the highest DF value, reaching 1.12, indicating that this group had higher requirements for the indoor thermal environment. The S-S group had the lowest DF (0.87), indicating that this group was more tolerant of thermal environments that deviated from thermal neutrality. Interestingly, the two groups that migrated, the N-S and the S-N subgroups exhibited different changing patterns in demand factors over time, which reflects different adaptation processes. During the first year after migration, the increment in the S-N subgroup's DF accounted for 85.7% of their total increment over three years, but the decrease in the N-S subgroup's DF only accounted for 29.7% of their total decrease over three years. The

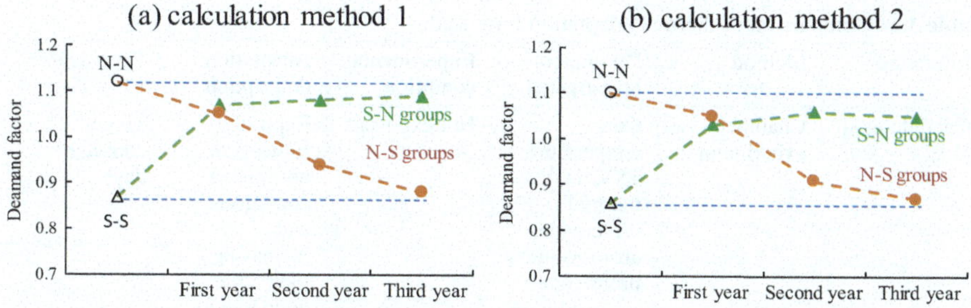

Fig. 3.8 Dynamics of demand factor (DF). **a** calculated from Eq. 3.5; **b** calculated from Eq. 3.6. *Note* That different DF calculation methods lead to slightly different DF values, but the overall trends in the two figures are quite similar. It is reasonable to believe either DF calculation method can well reflect the dynamic processes of thermal adaptation

S-N subgroup's DF increased rapidly during the first year and then remained steady, which means they adapted quickly to the Beijing indoor environments, and within a year were responding similarly to people who had lived in northern China for their entire lives. In comparison, the N-S subgroup's DF decreased gradually, implying a slow adaptation rate. This suggests that even after more than 3 years, this group still did not fully accept the cold indoor thermal environment in Shanghai.

The results shown in Figs. 3.7 and 3.8 indicate the thermal adaptation in buildings exhibit asymmetric dynamic trajectories—it is much quicker for occupants to raise their expectation and accept neutral indoor climate than to lower their expectation and acclimate to under-conditioned environments. It took the N-S subgroups 3 years to adapt to cold indoor temperatures while the S-N subgroup accepted the neutral and warm buildings in less than 1 year.

Table 3.6 summarizes some research mentioning the time-scale of thermal adaptation, physiologically or psychologically, or the combination of the two. The reported time scales varied from days to years. For those who migrated to new thermal environments, the adaptation process may start a few days or weeks after moving but may require months or years to be completely accomplished. These seemingly chaotic results once again proved the complexity of thermal adaptation, which may involve multiple layers. A full thermal adaptation seems to require a longer period of time, rather than being achieved in one stroke.

3.4.2 *Practical Implication*

The above analysis uses one year (or heating season) as the time scale to investigate the dynamic process of human thermal adaptation. In fact, if the time scale is shorter, these dynamic rules can be even stronger. To discuss this issue, we divided the Beijing survey into three stages: *Pre-Heating, In-Heating, and Post-Heating*, and then all respondents' thermal comfort evaluations during each stage were analyzed.

Table 3.6 Comparison of thermal adaptation time-scales

Literature	Method	Parameters investigated	Experimental condition	Adaptation description	Time-scale
Candido et al. [13]	Chamber experiment	Core temperature, TSV, TCV, thermal preference, air movement preference	Hot exposure	Expose subjects to hot-humid conditions and increasing their core temperature by exercise	Three consecutive days
Van der lans et al. [19]		Brown adipose tissue, energy expenditure, TSV, TCV, etc.	Cold exposure	10 consecutive days cold protocol	10 consecutive days
Luo et al. [20]		Skin blood flow, skin temperature, TSV, TCV, etc.	Cold exposure	From cold indoors to neutral-warm indoors	From 1 month to 1 year
Wijayanto et al. [21]		Rectal temperature, skin temperature, sweat rate	Hot exposure	From tropical area to temperate area	1–2 years
Yu et al. [11]		Heat stress protein, skin temperature	Hot exposure	From natural-ventilated building to air-conditioned building	>2 years
Liu et al. [14]	Field study	TSV, TCV, humidity sensation	Hot exposure	From mild climate zone to hot and humid climate zone	<8 days
Cao et al. [22]		TSV, TCV, clothing insulation	Cold exposure	From cold indoors to neutral-warm indoors	A heating season (~5 months)
Present study		TSV, TCV, clothing insulation	Cold exposure	From cold indoors to neutral-warm indoors	<1 year
				From neutral-warm indoors to cold indoors	~3 years

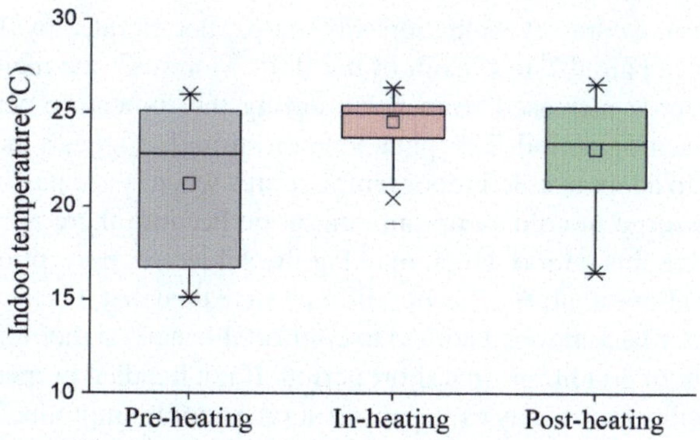

Fig. 3.9 Indoor temperature distribution at different heating stages

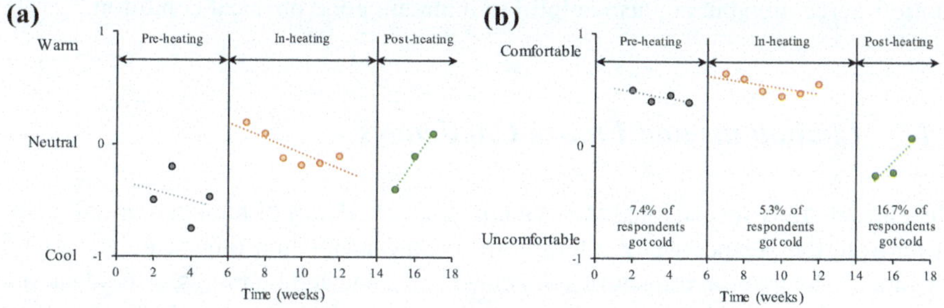

Fig. 3.10 Weekly average **a** thermal sensation vote and **b** thermal comfort vote. *Note* To make the various stages of the subjective thermal sensation vote comparable, the same correction method was applied as in Sect. 3.4, to obtain the actual thermal sensation vote TSV'

Figure 3.9 shows the indoor temperature distribution during different heating stages. Generally, the room temperature before heating is lower and distributed widely. During the heating season, the room temperature is the highest and is concentrated in a narrow range. Once the heating was suspended, the room temperature decreased slightly.

Figure 3.10a shows the weekly average thermal sensation vote during different heating stages. The preheating stage had relatively lower thermal sensation values and showed a downward trend. When the space heating started, subjects' thermal sensations were raised by 0.83 TSV scale units with a significance paired *T*-test result of $p < 0.01$. Similarly, when space heating was suspended, the corresponding thermal sensation dropped by 0.42 scale units with a paired T-test of $p < 0.05$.

Figure 3.10b presents weekly average thermal comfort votes. Respondents expressed that they were comfortable during the preheating period even though the ambient temperature was cold. After starting space heating in week 6, the subjects' thermal comfort votes improved even more, and remained steady at a fairly high level during the entire heating season. Once the heating season was suddenly stopped in week 15, the thermal comfort evaluation was sharply deteriorated by 0.87 scale units with a significance paired *T*-test result of $p < 0.01$. Moreover, the incidence of common cold symptoms increased from 5.3% during the in-heating period to 16.7% during the post heating period. This phenomenon strongly suggests that it is not easy for respondents to adapt to cold indoor temperatures when space heating is suddenly suspended. The surge in cold symptoms might be because there are physiological adaptations during this period which may lag the subjective perception changes.

Based on the above analysis, it is not difficult to realize that it is more challenging for those who have become accustomed to comfortable neutral thermal conditions to adapt to cold indoor conditions in a short period. If not handled properly, the sudden drop in indoor temperature may even cause a surge in cold symptoms. Therefore, it is essential to consider the dynamic characteristics of human thermal adaptation when planning air conditioning or space heating strategies. Gradual temperature changes rather than sudden temperature steps can not only help to improve occupants' thermal comfort perception but are also helpful to maintain good physical condition.

3.4.3 Limitations and Future Challenges

The current study investigated the dynamic characteristics of human thermal adaptation from the perspective of a long-term (years) adaptation time scale and found asymmetric adaptation trajectories in heated and unheated buildings. A field survey is often accompanied by confounding factors, making it difficult to clarify the effect of each factor. Generally, statistical analysis of large sample surveys is suitable to describe the phenomenon and summarize overall rules but can rarely explain the mechanism behind it. Given the limited sample size in the current study, we are unsure whether the sample size is sufficient. In future studies, with larger sample sizes should be conducted, and well-controlled comparative laboratory experiments should examine the psychological and physiological bases for human thermal adaptation. Additionally, the phenomena observed in Fig. 3.10 need further investigation, especially the common cold symptoms. More dedicated field studies and laboratory experiments should be designed to understand the underlying principles behind these observations.

3.5 Conclusions

Through two comparative thermal comfort studies in Shanghai and Beijing, this paper investigates the dynamic process of thermal adaptation to heated and unheated building environments. The main findings are as follows.

(1) Building occupants' thermal adaptation exhibits asymmetric trajectories—it is much quicker for occupants to raise their expectations and accept a neutral indoor climate than to lower their expectations and acclimate to under-conditioned environments. In terms of adaptation time, it took the northern groups 3 years to adapt to cold, unheated indoor conditions, while the southern groups accepted the neutral and warm heated indoor environments in less than 1 year.
(2) Although there are many ways to define the metrics of "demand factor", its basic concept can be used to track the dynamic processes of thermal adaptation and the transformation of thermal comfort expectations. In this study, we found the "demand factor" can well reflect the asymmetric trajectories of thermal adaptation.
(3) When delivering building thermal environments, the dynamic characteristics of human thermal adaptation should be respected. Using space heating as an example, if the dynamics of thermal adaptation can be considered and a sudden increase or decrease in space heating temperatures avoided, occupants can not only benefit from an improvement in thermal comfort but also become less likely to experience the cold symptom.

References

1. Deuble MP, de Dear RJ (2012) Mixed-mode buildings: a double standard in occupants' comfort expectation. Build Environ 54:53–60
2. Luo M, Cao B, Damiens J, Lin B, Ouyang Q et al (2015) Evaluating thermal comfort in mixed-mode buildings: A field study in a subtropical climate. Build Environ 88:46–54
3. Rupp R, de Dear R, Ghisi E (2018) Field study of mixed-mode office buildings in Southern Brazil using an adaptive thermal comfort framework. Energy Build 158:1475–1486
4. Candido C, de Dear R, Lamberts R, Bittencourt L (2010) Cooling exposure in hot humid climates: are occupants 'addicted'? Archit Sci Rev 53(1):59–64
5. Ghahramani A, Zhang K, Dutta K et al (2016) Energy savings from temperature setpoints and dead band: quantifying the influence of building and system properties on savings. Appl Energy 165:930–942
6. de Dear R (1998) Global database of thermal comfort field experiments. ASHRAE Trans 104:1141–1152
7. de Dear R, Brager G (2001) The adaptive model of thermal comfort and energy conservation in the built environment. Int J Biometeorol 45:100–108
8. Carlucci S, Bai L, de Dear R, Yang L (2018) Review of adaptive thermal comfort models in built environmental regulatory documents. Build Environ 137:73–89
9. Zaki S, Damiati S, Rijal H, Hagishima A, Razak A (2017) Adaptive thermal comfort in university classrooms in Malaysia and Japan. Build Environ 122:294–306
10. Ning H, Wang Z, Zhang X, Ji Y (2016) Adaptive thermal comfort in university dormitories in the severe cold area of China. Build Environ 99:161–169
11. Yu J, Ouyang Q, Zhu Y et al (2012) A comparison of the thermal adaptability of people accustomed to air conditioned environments and naturally ventilated environments. Indoor Air 22:110–118
12. van der Lans A, Hoeks J, Brans B et al (2013) Cold acclimation recruits human brown fat and increases no-shivering thermogenesis. J Clin Investig 123(8):3395–3403

13. Candido C, de Dear R, Ohba M (2012) Effects of artificially induced heat acclimatization on subjects' thermal and air movement preferences. Build Environ 49:251–258
14. Liu Y, Yu D, Cong S et al (2017) A tracked field study of thermal adaptation during a short-term migration between cold and hot-summer and warm-winter areas of China. Build Environ 124:90–103
15. Nicol J, Humphreys M (2002) Adaptive thermal comfort and sustainable thermal standards for buildings. Energy Build 34(6):563–572
16. Humphreys MA (1976) Field studies of thermal comfort compared and applied. J Inst Heat Ventilating Eng 44(1):5–27
17. ISO EN ISO 8996 (2004) Ergonomics of the thermal environment—determination of metabolic rate. International organization for Standardization, Geneva
18. Foldvary V, Cheung T, Zhang H et al (2018) Development of the ASHRAE global thermal comfort database II. Build Environ 142:502–512
19. van Marken Lichtenbelt W, Vanhommerig J, Smelders N et al (2009) Cold activated brown adipose tissue in healthy men. N Engl J Med 360(15):1500–1508
20. Luo M, Ji W, Cao B et al (2016) Indoor climate and thermal physiological adaptations: evidences from migrants with different cold indoor exposures. Build Environ 98:30–38
21. Wijayanto T, Toramoto S, Wakabayashi H, Tochihara Y (2012) Effects of duration of stay in temperate area on thermoregulatory responses to passive heat exposure in tropical south-east Asian males residing in Japan. J Physiol Anthropol 31(1):1–10
22. Cao B, Zhu Y, Ouyang Q, Zhou X, Huang L (2011) Field study on human thermal comfort and thermal adaptability during the summer and winter in Beijing. Energy Build 43(5):1051–1056

Chapter 4
Indoor Climate and Physiological Acclimation

4.1 Statement of the Problem

4.1.1 Indoor Climate and Physiological Adaptation

Although existing adaptive comfort studies have shed useful light on how occupants adapt to the surrounding thermal environment [1–4], the indoor climate change since the appearance of HVAC technologies begs for a question as to whether long-term indoor thermal exposures will influence occupants' thermal response, especially with regard to their physiological response. The significance of this seemingly simple question cannot be overstated as they are closely related to human well-being and are vital for shaping the roadmap of the future indoor climate.

To date, researchers have not reached a consensus on whether indoor thermal history has a greater impact on thermal adaptation than outdoor thermal experience, or vice versa. Although it is a small body of evidence, some relevant studies can be introduced here. Johnson et al. [5] reported in their literature review that reduced exposure to cold could be expected to have a dual impact on body weight. More time spent in the thermal neutral zone not only reduce the frequency of occasions on which cold-induced energy expenditure is initiated but also inevitably lead to reduced thermogenic capacity. Kingma et al. [6] defined a thermoneutral zone including body core, skin, and ambient temperature. The result suggests that human behavioral adjustment may be initiated already before the boundaries of the thermoneutral zone are reached. Yu et al. [7] conducted a climate chamber experiment with a group of subjects acclimated to air-conditioned environments (AC group) and a control group acclimated to naturally ventilated environments (NV group). The results suggested that in hot exposures, the NV group had a stronger capability for physiological thermoregulation than AC group. Van der lans et al. [8] found that a 10-day cold acclimation protocol can increase brown adipose tissue activity in parallel with an increase in non-shivering thermogenesis in humans. After 10-day's consecutive cold

M. Luo, *The Dynamics and Mechanism of Human Thermal Adaptation in Building Environment*, Springer Theses,
https://doi.org/10.1007/978-981-15-1165-3_4

acclimation (15–16 °C), subjects experienced less cold perception, felt less uncomfortable and had fewer shivering reports during the cold exposures. All these studies collectively support the correlations between indoor climate and human thermal adaptation from different angles, especially with regard to short-term or long-term physiological acclimations. However, more specific evidence are needed to elaborating the effects of physiological acclimations on building occupants' thermal adaptation.

4.1.2 Objectives of This Chapter

To better understand the physiological acclimation phenomenon and its effects on thermal adaptation, this chapter tries to answer the following questions. (1) Does physiological acclimatization exist? What kind of index can reflect it? (2) How much is the weighting of physiological acclimatization in overall thermal adaptation? How does it change with adapting time? (3) How to describe physiological acclimating and incorporate it into a more comprehensive adaptive model?

4.2 Method

4.2.1 Overall Experimental Plan

To answer the above three questions, we designed three corresponding climate chamber experiments.

First, we answered the question as to 'whether there exists physiological acclimation and how to represent it.' An experiment was conducted to compare the cold adaptation capability of people from the south and north China. Different physiological indexes were screened to reflect subjects' physiological acclimation. *Second*, we investigated the adapting process, and the weighting factor of physiological acclimation. *Third*, we tried to quantify the physiological acclimation effects on building occupant's thermal adaptation. To explain these experiments more clearly, we will detailly introduce the first-stage experimental design and then supplement the following experiment procedures in following sections.

4.2.2 Subjects Information

The generalization of wintertime indoor temperature difference between northern and southern China (as described in Chap. 2) offered a natural opportunity for comparative research on physiological acclimation. Two subject groups with different indoor

Table 4.1 Subjects' profile

Group	Description	Sample size	Age (year)	Height (cm)	Weight (kg)	BMI
N-N	Grow up in cold climate zone with district space heating	15	20.5 ± 1.3	176.3 ± 6.2	65.6 ± 4.9	21.1 ± 1.1
S-N	Grow up in hot summer and cold winter climate zone without district space heating	15	18.8 ± 0.7	171.9 ± 3.3	63.3 ± 4.5	21.6 ± 1.2

thermal histories were recruited on the basis of their wintertime indoor thermal exposures (Table 4.1). N-N group consists of subjects who had spent their entire life in the heated northern region with neutral-to-warm wintertime indoor climates (with space heating both at home and school dormitory). S-N group consists of subjects selected from those who grew up in the hot summer and cold winter climate zone without space heating facilities (no space heating both at home and school dormitory) and recently moved to the north. Each group contains 15 healthy volunteers who did not know much about thermal comfort and had no smoking habit. All the subject were asked to do some physiological measurements and subjective response in a climate chamber. To exclude the influence of gender, age, weight and etc., all the subjects were male college students. Table 4.1 summarizes their profiles. Although the N-N subjects had slightly higher heights, their BMI indexes were controlled in the normal range and were close to S-N group.

4.2.3 Experimental Protocol

Five experimental conditions (26/24/21/18/16 °C) were created in an artificial climate chamber, where temperature and humidity can be controlled strictly with precisions of ±0.2 °C and ±5%, respectively. As shown in Fig. 4.1, the experiment was carried out in a climate chamber with two rooms that can be controlled independently by two air handling units (AHU). The heater, humidifier, and cooling coil in AHUs are controlled by a computer program that minimizes the deviation between the measured thermal parameters and their set-points. The precisions of air temperature and relative humidity control are ±0.2 °C and ±5%, respectively. In the chamber, conditioned air is supplied from a ceiling perforated panel and exhausted through a raised floor. The

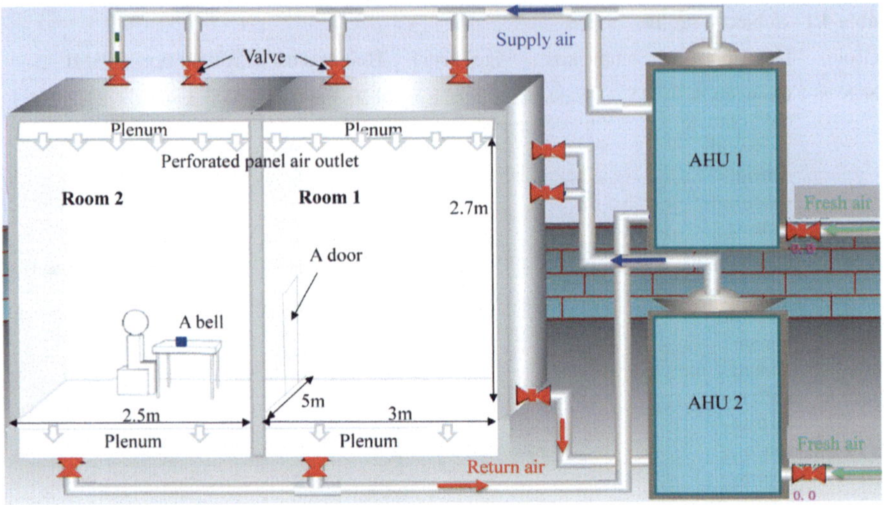

Fig. 4.1 Schematic diagram of the climate chamber

background air velocity was validated with omnidirectional anemometers (SWEMA 03+) at three heights (1.1, 0.6, and 0.1 m above the floor). The measured background air velocities (<0.1 m/s) were ensured lower than the sensible threshold (0.2 m/s), which has no significant impacts on subjects' thermal comfort perception. Twenty-four thermocouple thermometers with ±0.2 °C precision were distributed in three height levels (1.5, 1.0, and 0.5 m above the floor) to monitor the uniformity and stability of the thermal environment. The measured temperatures were utilized as feedback values for the control program. It is noteworthy that there is a door between the two independent rooms. Subjects can move from one room to another when it is necessary.

Figure 4.2 shows the detailed experiment protocol. When the experiment began, subjects had 10 min to calm down, change clothes, learn how to vote, as well as to wear physiological sensors. All the subjects were required to wear uniform clothes (underwear, T-shirt, shorts, socks, and sports shoes) with total clothing insulation of 0.42 clo. The reason that we chose light dress mainly lies in the convenience to wear physiological sensors. Then, the subjects were asked to stay in a neutral temperature (26 °C) for 30 min. Both physiological measurement and subjective questionnaire were started from the beginning of this phase. Finally, the subjects moved to other cold cases (24/21/18/16 °C) for 60 min.

The lower part of Fig. 4.2 illustrates the logic behind the selection of the experimental conditions. Before determining the specific experimental temperatures, PMV values were calculated in advance as a reference. 26 °C is close to thermally neutral, 24 °C is slightly cool, 21 °C is in the cool zone, 18 °C is cold discomfort, and 16 °C is intolerable cold. The five selected temperatures ranged from thermally neutral to intolerable cold.

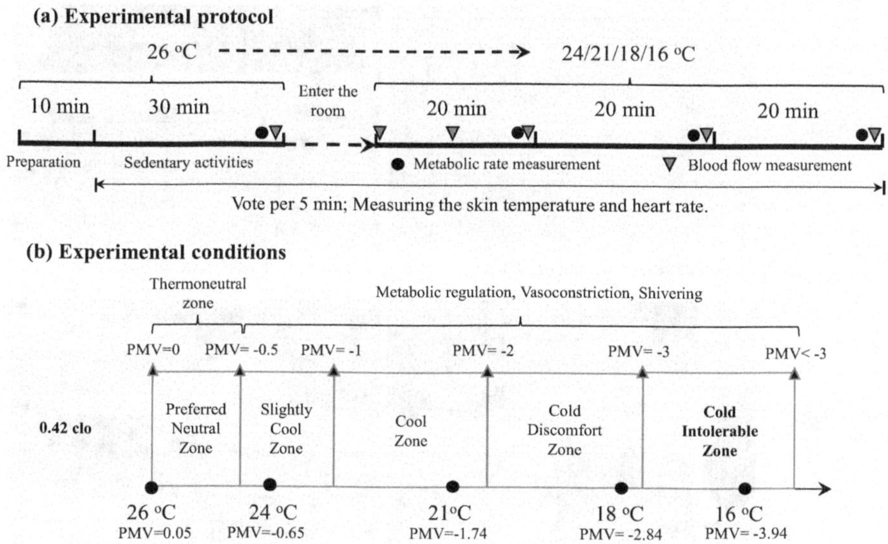

Fig. 4.2 Experimental protocol

To ensure the validity of experimental data, subjects were reminded to avoid caffeine, alcohol, smoking and strenuous activities, and to sleep well in the day prior to the experiment. At the beginning of recruitment, voluntary subjects were informed that there would have some cold exposures and that they were free to interrupt the experiment at any time if they felt unwell. For each subject, the temperature cases were ordered randomly. The whole experiments were conducted in Beijing, from late September to early October. Typically, this season is the best period in Beijing. The outdoor temperatures were comfortable, ranging from 18 to 25 °C. As all the S-N subjects moved to northern China in early September, they haven't experienced cold exposure or space heating in northern China.

4.2.4 Physiological Measurements

During the experiment, physiological parameters such as metabolic rate (measured by Vmax Encore, Sensor Medics, USA), skin surface blood flow rate (measured by PeriFlux 5001, Perimed, Sweden), skin temperature (measured by Vitalsense Dermal Patch) and heart rate (measured by Equital) were collected. The test procedure was shown in Fig. 4.2 and the test scene was shown in Fig. 4.3.

The skin temperatures were measured during the entire 90 min exposure through wireless sensors (with a precision of 0.1 °C) attached at nine body areas: forehead, chest, back, upper arm, forearm, hand back, thigh, calf, and ankle. The mean skin temperature was calculated using the following equation [9]:

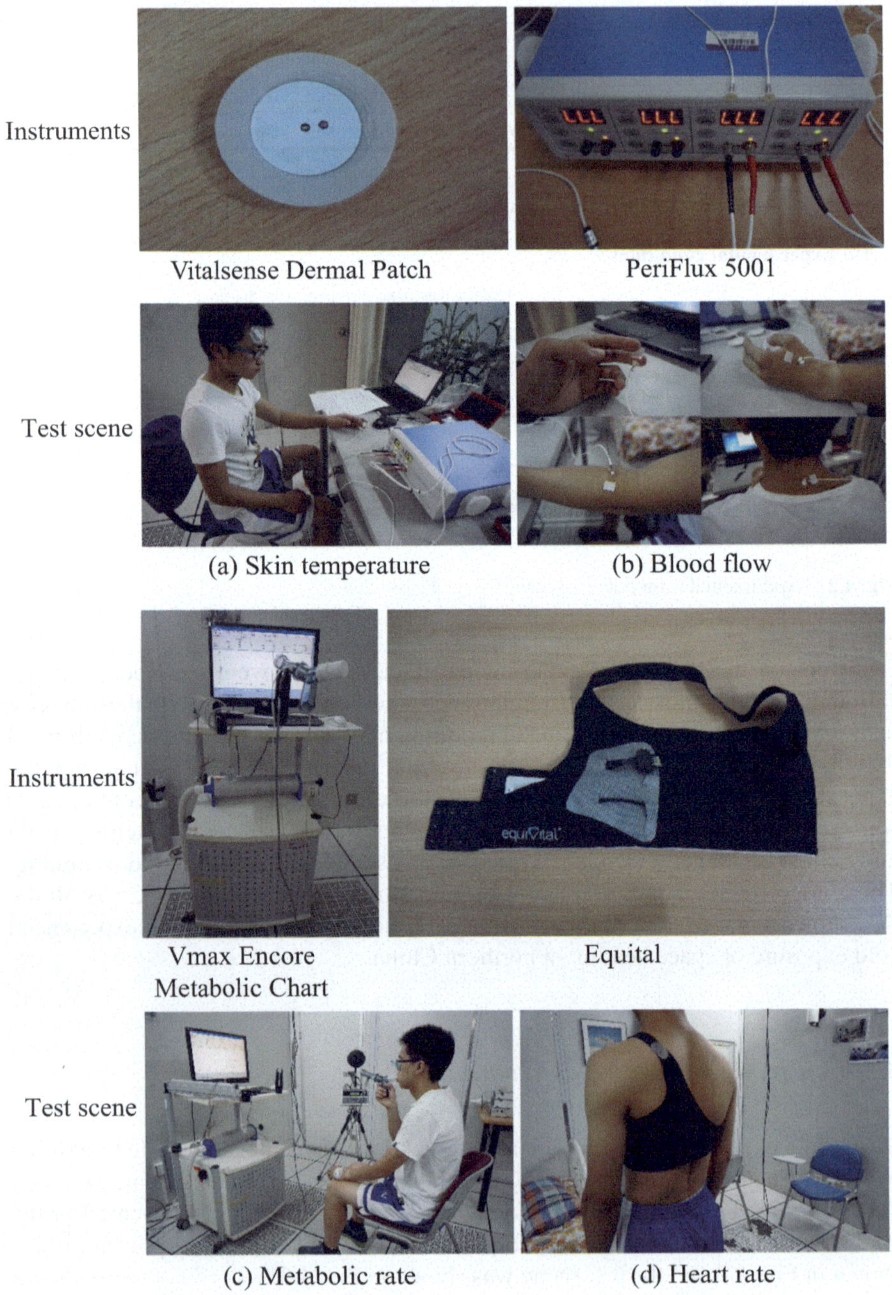

Fig. 4.3 Experimental device and test scene

$$T_{ms} = 0.07T_{forehead} + 0.18T_{chest} + 0.18T_{back} + 0.07T_{forearm} + 0.07T_{upper\ arm} + 0.05T_{hand\ back} + 0.19T_{thigh} + 0.13T_{calf} + 0.06T_{ankle} \quad (4.1)$$

The blood flow was continuously measured throughout the 90-min experiment, but only the average value in 2 min before and after the time shown in Fig. 4.2 was considered. In order to reduce the repeated work of blood flow measurement, a pre-experiment was designed before the formal experiment, in which that skin surface blood flow at seven body areas (left hand back, left forearm, neck back, forehead, left thigh, left ankle, and left middle fingertip) were measured in two temperature cases (26 and 18 °C). Based on preliminary test results, the skin surface blood flow at left middle fingertip was chosen as the representative parameter. Hence, in the formal experiment, only blood flow at the middle fingertip surface was measured.

The metabolic rate was measured before the end of the adaptation phase and every 20 min thereafter. The length of each measurement was 5 min. The oxygen concentration and carbon dioxide concentration changes during the respiration were measured, and then the metabolic rate was calculated according to Eq. (4.2) [10]:

$$M = \frac{21(0.23RQ + 0.77)Q_{O_2}}{A_d} \quad (4.2)$$

where M denotes the metabolic rate (W/m^2), RQ is the respiratory quotient, Q_{CO_2} is human body CO_2 production flowrate (mL/s),[1] Q_{O_2} is the oxygen flowrate consumed by human body (mL/s), A_d is Dubois body surface area (m^2) calculated by Eq. (4.3).

$$A_d = 0.202H^{0.725}WE^{0.425} \quad (4.3)$$

where H denotes height (m) and W means weight (kg).

Additionally, the heart rate was monitored during the whole 90 min exposure, as shown in Fig. 4.3d.

4.2.5 *Subjective Questionnaire*

In addition to physiological measurements, subjects were asked to vote subjective questionnaire every 5 min during the whole experimental period. The questionnaire includes the conventional seven-point thermal sensation vote (TSV), a seven-point thermal comfort vote (TCV), self-reported shivering rate, as well as other miscellaneous information. The TSV and TCV scales with their corresponding numerical values are depicted in Table 4.2. For the question regarding to shivering rate, the

[1] The carbon dioxide volumetric flow rate produced by human body and the oxygen volumetric flow rate consumed need to be converted into the flow rate under standard conditions (0 °C, 101.3 kPa).

Table 4.2 Thermal comfort index scale

Scale	TSV	TCV	Shivering
3	Hot	Very comfortable	–
2	Warm	Comfortable	–
1	Slightly warm	Just comfortable	Yes
0	Neutral	unclear	–
−1	Slightly cool	Just uncomfortable	None
−2	Cool	Uncomfortable	–
−3	Cold	Very uncomfortable	–

subjects could either vote ‘yes’ to indicate that they shivered during the experiment or vote ‘none’ to indicate no shiver happened. Then, the shivering rate was calculated as the percentage of subjects who reported shivering in a specific temperature case.

To compare the data from different groups, Independent Samples *T*-Test or Paired-Samples *T*-Test were utilized. When the test result shows the significance of difference, it is labeled as ‘$p < 0.05$’ or ‘$p < 0.01$’, otherwise, ‘$p > 0.05$’ was labeled.

4.3 Representative Physiological Parameters

4.3.1 Basic Characteristics of Physiological Response

Table 4.3 summarizes the measured thermal parameters and the calculated PMV values in different experimental conditions. The thermal conditions were effectively controlled in accordance with the designed set-points. The variance in the monitored thermal parameter could only result in less than 0.1 unit scale of PMV.

Figure 4.4 illustrates the results of skin surface blood flow measurement conducted in 26 and 18 °C conditions. Firstly, the skin surface blood flow decreased as the ambient temperature dropped. Secondly, the blood flow at the middle fingertip was higher than other body parts, and its drop with temperature decreasing was more significant. Therefore, the blood flow at left middle fingertip was chosen as a represent in following formal tests.

Figure 4.5 depicts the results of skin temperature measurement in the thermally neutral condition of 26 °C. The skin temperatures varied from one body part to another. Large individual differences existed in each body part as well. In the following analysis section, the mean skin temperature [calculated according to Eq. (4.1)] was adopted.

Table 4.3 Statistical summary of measured thermal parameters

Temperature cases	26 °C		24 °C		21 °C		18 °C		16 °C	
	Mean	Std.	Mean	Std.	Mean	Std.	Mean	Std.	Mean	Std.
t_a (°C)	26.1	0.23	23.9	0.18	20.9	0.13	18.1	0.21	16.1	0.34
MRT (°C)	25.9	0.11	24.0	0.13	20.9	0.10	18.2	0.13	16.2	0.26
t_{op} (°C)	26.0	0.14	24.0	0.12	21	0.15	18.1	0.13	16.1	0.29
ν (m/s)	0.06	0.03	0.04	0.02	0.07	0.02	0.10	0.04	0.09	0.05
RH (%)	52.1	4.2	47.5	3.5	52.5	4.9	53.9	3.6	51.7	4.5
PMV	0.06	0.13	−0.66	0.12	−1.75	0.05	−2.86	0.09	−3.93	0.11

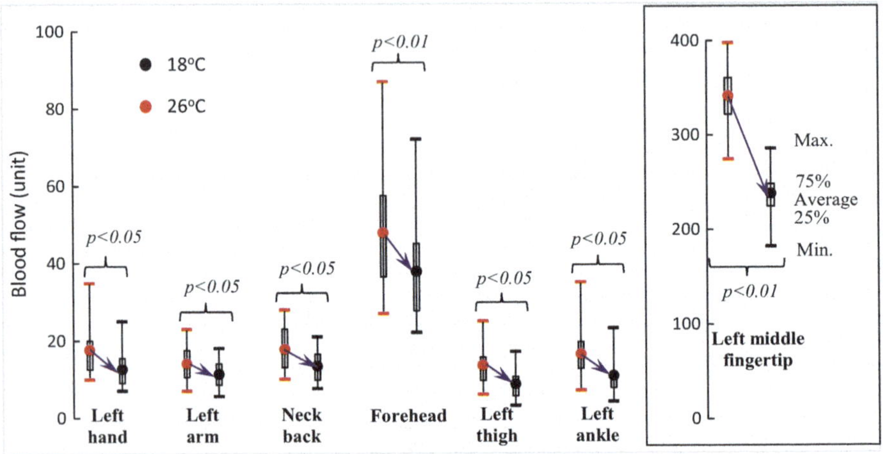

Fig. 4.4 Skin surface blood flow at different body areas

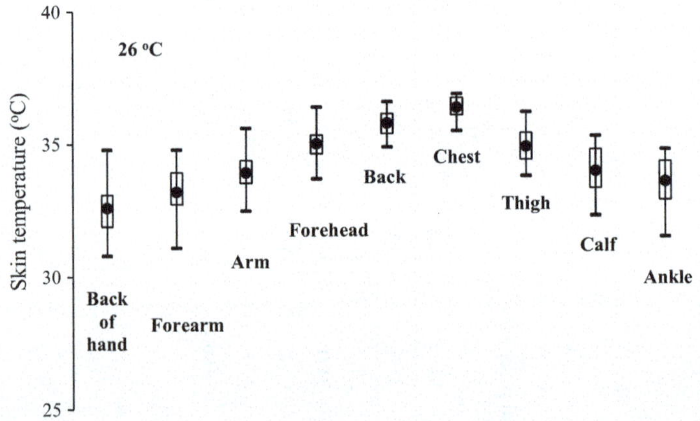

Fig. 4.5 Skin temperature at different body areas

4.3.2 Selecting Representative Physiological Parameters

Figure 4.6 compares the physiological response between N-N group and S-N group. The data includes results from the steady-state of each temperature case. In general, when the temperature dropped from 26 to 16 °C, the average metabolic rate increased nearly 14%, the mean skin temperature dropped from 33.8 to 30.9 °C, the average blood flow at left middle fingertip decreased from 350 unit to 213 unit, while the heart rate did not show significant changing.

Further comparing the differences between the two subject groups, it can be seen that no significant difference exists in metabolic rate and heat rate. However, the

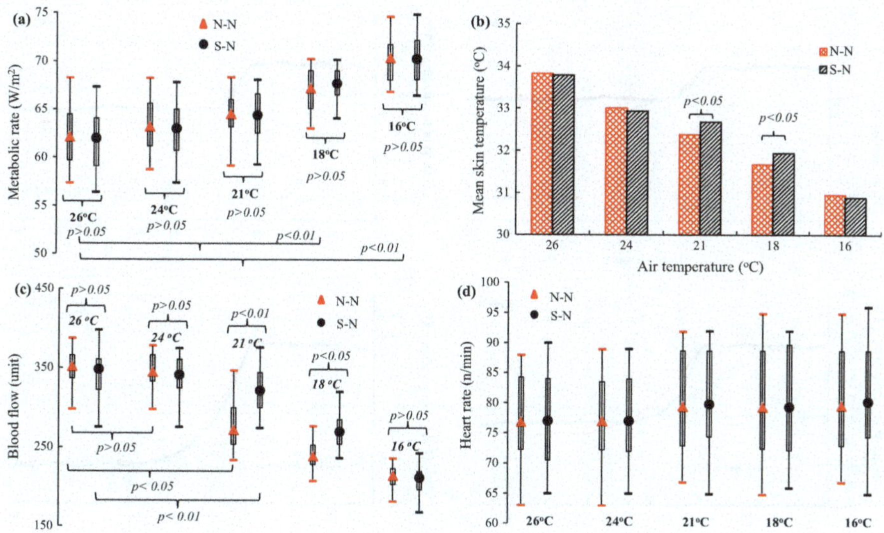

Fig. 4.6 Physiological response differences between N-N group and S-N group: **a** metabolic rate, **b** mean skin temperature, **c** blood flow at the left middle fingertip, **d** heart rate

mean skin temperature and blood flow showed some differences between these two groups. Especially in mildly cold conditions (21 and 18 °C), S-N subject group had higher mean skin temperature and higher blood flow. In conditions close to thermal neutrality (26 and 24 °C) and intolerable cold temperature (16 °C), no significant differences were found.

Figure 4.7 illustrates the changing process of mean skin temperature in different experimental cases. Mean skin temperature decreased with the drop in ambient temperatures. In 21 and 18 °C cases, N-N subject group had slower skin temperature decreasing rate during the adjusting process and lower skin temperature after reaching steady state. Similarly, the differences only exist in 21 and 18 °C cases. No significant differences were observed in other temperature cases.

Figure 4.8 compares the subjective thermal sensations, thermal comfort, and self-reported shivering rate between N-N group and S-N group. The data includes results from the steady-state of each experimental case. Compared with N-N group, S-N group felt less cold and less uncomfortable when they were exposed to mild cold conditions (21 and 18 °C). In other neutral temperatures (26 and 24 °C) and intolerable cold case (16 °C), no significant differences between these two groups were found.

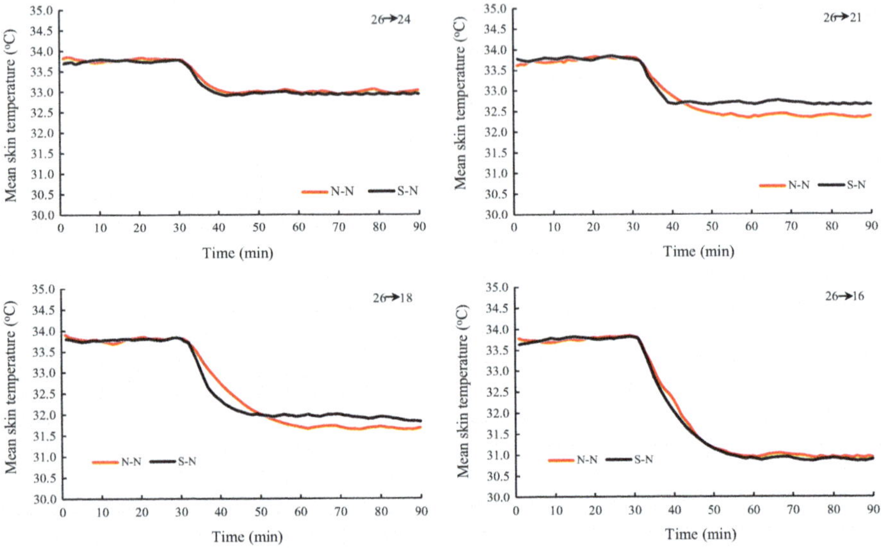

Fig. 4.7 Comparison of mean skin temperature in different experimental conditions

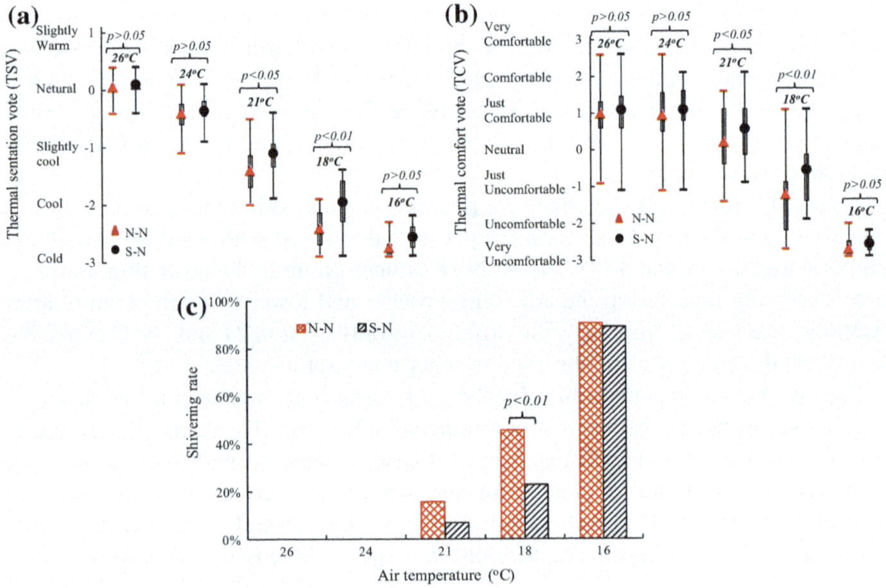

Fig. 4.8 Subjective perception differences between N-N group and S-N group: **a** thermal sensation vote, **b** thermal comfort vote, **c** shivering rate

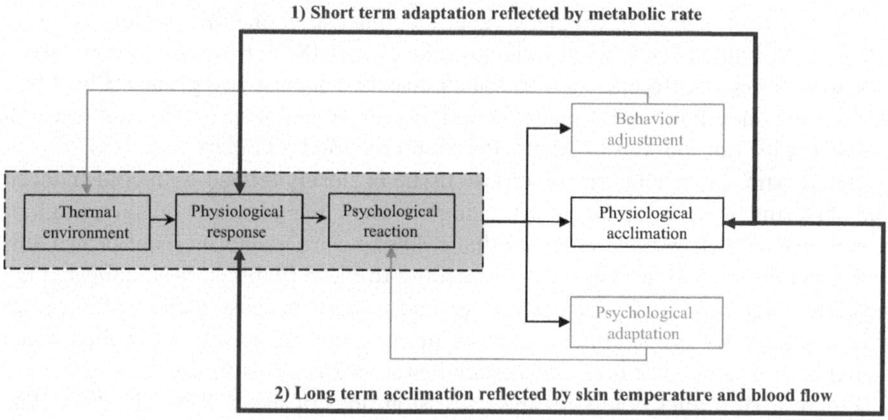

Fig. 4.9 Timescales of physiological acclimation

4.3.3 Physiological Acclimation to Cold Exposure

4.3.3.1 The Framework of Physiological Acclimation

According to the results shown in Figs. 4.6 and 4.7, although the metabolic rate cannot reflect the difference caused by long-term thermal adaptation, it varies significantly with the thermal environment and can be regarded as an important indicator of short-term physiological regulation. Meanwhile, the skin blood flow and average skin temperature can reflect the difference caused by long-term thermal adaptation under moderate to cold (21–18 °C) conditions. Therefore, the effects of physiological acclimatization on human thermal balance and subjective feelings can be described from two-time scales, as shown in Fig. 4.9. On one hand, physiological regulation in the short term can be reflected by metabolic rate difference. On the other hand, long-term physiological acclimatization can be reflected by skin temperature and blood flow.

4.3.3.2 Physiological Acclimation and Indoor Thermal Experience

Figures 4.4, 4.5, 4.6 and 4.7 illustrates the subjects' physiological adaptations to cold. All these results collectively point to a conclusion that the human body itself can adapt to cold environment through thermoregulation to some extent. When subjects were exposed to cold ambient temperatures, both skin surface blood flow, and skin temperature would decrease, thus preventing too much heat dissipation of human body. Figure 4.5 also sheds some light on the question 'how long human body can tolerate the cold conditions'. It can be seen that the mean skin temperature was able to reach a steady state in around 30 min and then maintained at a stable level for another 30 min.

It is noteworthy that N-N subjects reported higher shivering rate (which is closely related with human body heat production) in 21 and 18 °C (Fig. 4.8c), while there was no significant difference in metabolic rate between these two groups (Fig. 4.6a). What is the underlying mechanism behind this phenomenon? Some recent research conducted by van Marken Lichtenbelt's group reported a kind of fat called "brown adipose tissue" [11]. The brown adipose tissue is closely related with cold-induced non-shivering thermoregulation (thermogenesis in the absence of shivering) of human body. Their results indicated that cold exposure would increase brown adipose tissue mass and activity [12]. According to these findings, S-N subjects may have increased brown adipose tissue mass and activity, because they lived in colder wintertime indoor temperatures. And this might be the reason why S-N subjects had higher shivering rate but the same metabolic rate with N-N subjects.

Figure 4.10 further compares the regression lines of mean TSV and PMV. The PMV index was calculated by using the measured thermal parameters (air temperature, mean radiant temperature, air velocity, and relative humidity), clothing insulation (0.42 clo) and assumed metabolic rate for sedentary activity (1.1 met). rPMV was revised PMV by using the measured metabolic rate in each temperature case. All the data dots of TSV are the average values for all subjects during the last 20 min of each temperature case. Some interesting phenomena should be noteworthy.

Firstly, the slope of TSV regression line for S-N group is slighter than that for N-N group, which indicates that S-N subjects felt less cold than N-N subjects did. This result is consistent with the physiological test introduced above. Normally, a colder TSV correspondents to lower skin temperature, lower skin surface blood flow rate, higher metabolic rate, and higher self-reported shivering rate. All these results collectively point to the conclusion that subjects with colder indoor exposures had a slighter physiological response and felt less uncomfortable in mildly cold conditions.

Secondly, the "scissors difference" (different slopes of regression lines) between PMV and TSVs indicates that PMV overestimated subjects' cold sensations. This may be resulted from the inaccurately assumed metabolic rate (1.1 met). If the actual measured metabolic rates were utilized, the rPMV was quite close to the TSV from

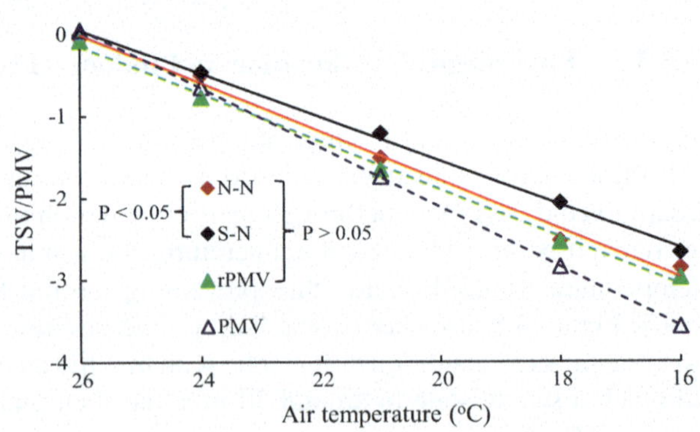

Fig. 4.10 Comparison of TSVs, PMV and rPMV

N-N group. This phenomenon reminds us to consider the metabolic rate more carefully in future comfort studies. However, it should be noted that rPMV still overestimated the actual sensations of S-N group.

4.3.3.3 Individual Differences in Physiological and Subjective Responses

From Figs. 4.4, 4.5, 4.6, 4.7 and 4.8, wide distribution ranges can be observed in both physiological and subjective test results, which indicates there exist individual differences. To illustrate individual adaptive processes more clearly, Fig. 4.11 shows TSV and skin surface blood flow changes of each subject. It can be seen that individual differences happened in almost every temperature case. It seems like that thermal comfort can be achieved in a relatively wide range of parameters rather than an absolute value. This might be the reason why comfort evaluations are commonly treated as statistical percentage based on large samples, and thermal comfort studies usually adopt statistical analysis instead of merely looking at a single sample.

Certainly, the individual comfort response can also provide another angle to view thermal comfort. For instance, Fig. 4.11 sheds some light on the phenomenon 'why

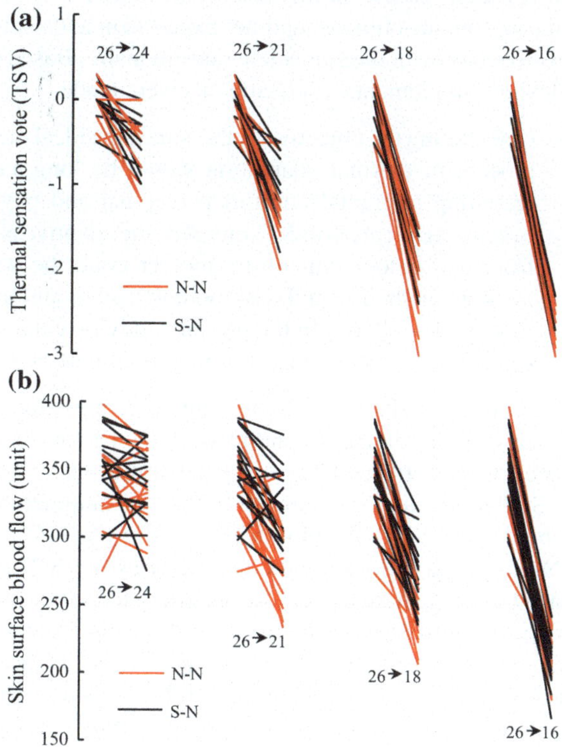

Fig. 4.11 Individual response in different temperatures (each line stands for one subject): **a** thermal sensation vote, **b** skin surface blood flow

the differences between S-N and N-N group were not obvious in neutral and intolerable cold cases'. It might be because 16 °C was far beyond the comfort range, almost all the subjects complained of cold discomfort. And 24 °C was still close to thermal neutrality, subjects' responses were still in chaos.

4.4 Long-Term Physiological Acclimation

In Sect. 4.3, we investigated the basic characteristics and representative indicators of human body physiological acclimation and discussed the individual differences of physiological acclimation and its relationship with indoor thermal experience. In this section, we continued to explore the physiological acclimation from a long-term view and focus on the changing process of physiological acclimation and its proportion in the overall thermal adaptation.

4.4.1 Experiment Supplement

The research method in this section is similar to that in the previous section. We will carry out the climate chamber experiment and analyze the physiological parameters together with the subjective questionnaire. Based on previous experiments, the following supplementary adjustments were made.

(1) Considering the objective of this study, the ideal method to examine the dynamic process of thermal adaptation would be long-term tracking studies, namely recording migrants' comfort perceptual and physiological response changes over a long-term period. In this way, the adaptive process of subjects who moved from one indoor climate to another could be well documented. But such a tracking method would cost too much labor and time. Therefore, an alternative method was chosen in the present study by comparing the thermal response of subjects who moved to a new indoor climate at different times.

According to the above logic, four subject groups were recruited (see Table 4.4 Subjects' profile in the second phase test). For the N-N group, subjects spent their entire life in northern China with neutral-to-warm wintertime space heating. For the S-N group, subjects grew up in the "hot summer and cold winter" climate zone without space heating and moved to Beijing before the space heating started. The S-N-1m group followed similar criteria as the S-N group, but they had 1-month space heating experience before attending the experiment. Likewise, for the S-N-1y group, subjects were similar to the S-N group, except they had lived in Beijing for one year before the test. Each group contains 15 healthy volunteers who did not know much about thermal comfort. Table 4.4 summarizes their profiles. To exclude the influence of gender, age, weight, etc., only healthy male college students were included. Prior to the experiment, all the subjects were reminded to avoid caffeine,

Table 4.4 Subjects' profile in the second phase test

Group	Sample size	Age (year)	Height (m)	Weight (kg)	BMI
N-N	15	20.5 ± 1.3	176.3 ± 6.2	65.6 ± 4.9	21.1 ± 1.1
S-N	15	18.8 ± 0.7	171.9 ± 3.3	63.3 ± 4.5	21.6 ± 1.2
S-N-1m	15	18.8 ± 0.7	174.9 ± 3.3	63.3 ± 4.5	21.6 ± 1.2
S-N-1y	15	19.7 ± 0.5	172.9 ± 4.5	62.3 ± 3.8	21.6 ± 1.4

alcohol, smoking and strenuous activities, and to sleep well. Besides, all voluntary subjects were informed that they would experience some cold exposures and that they were free to interrupt the experiment at any time if they felt unwell or uncomfortable. Then they signed an inform consent.

(2) To choose proper experimental conditions, the PMV of each temperature case was calculated in advance. The 26 °C temperature is close to thermally neutral with PMV of 0.05; 24 °C is slightly cool with PMV of −0.65; 21 °C is in the cool zone with PMV of −1.74; 18 °C is cold discomfort with PMV of −2.84; and 16 °C is intolerably cold with PMV of −3.94. According to the preliminary study presented in Sect. 4.3, we found no significant difference between N-N group and S-N group for the 24 and 16 °C cases. The reason might be because 24 °C was too close to thermal neutrality, so the subjects' physiological response was not strong. Conversely, 16 °C was far beyond the comfort range, which imposed a very strong cold stimulus to all subjects. Considering the fact that cold temperatures in building environments are normally no worse than moderately cold, we, therefore, decided to exclude 24 and 16 °C cases from the present study.

4.4.2 *Basic Characteristics of Physiological Acclimation*

Table 4.5 summarizes the statistical result of measured physiological parameters. All the data in this table were composed of results from the last ten minutes of each test case. To compare the data from group N-N and the other three groups, independent *T*-test was utilized. The test result is labeled as '*' when $p < 0.05$ and '**' when $p < 0.01$, which means there exists a statistically significant difference between the tested groups. Otherwise, 'o' was labeled, indicating no significant difference (i.e., $p > 0.05$).

For metabolic rate and heart rate, Table 4.5 shows no difference between N-N group and other migrated groups. But it is noteworthy that temperature conditions can greatly influence human metabolism. For instance, the average metabolic rate at 18 °C was nearly 9% higher than that at 26 °C. Analogous effects were observed in

Table 4.5 Statistical summary of measured physiological parameters

Categories			26 °C				21 °C				18 °C			
			S-N	S-N-1m	S-N-1y	N-N	S-N	S-N-1m	S-N-1y	N-N	S-N	S-N-1m	S-N-1y	N-N
Metabolic rate (W/m^2)	Statistic result	Min.	55.9	55.4	56.4	56.6	59.3	59.2	59.4	59.1	63.1	62.8	61.3	64.1
		25%	59.6	58.0	58.5	58.3	62.3	62.4	62.7	62.3	65.1	64.8	64.1	66.6
		Mean	60.8	60.9	60.8	61.0	64.1	64.0	63.8	64.0	67.2	67.0	67.5	67.7
		75%	62.0	63.9	63.4	63.1	66.3	66.1	65.5	65.9	69.3	69.8	68.3	69.1
		Max.	66.7	66.1	67.3	66.3	68.4	67.3	68.5	68.1	70.3	71.3	70.6	70.2
		Median	60.6	61.4	60.7	59.7	64.7	63.9	65.0	64.2	67.3	66.9	67.1	67.5
		Std.	3.18	3.04	2.91	3.34	2.87	2.75	2.81	2.81	2.21	1.69	2.21	1.69
	Significance of difference	N-N	o	o	o	–	o	o	o	–	o	o	o	–
Heart rate (n/min)	Statistic result	Min.	63.0	65.0	63.0	65.0	67.0	69.0	67.0	65.0	66.0	67.0	69.0	67.0
		25%	70.5	69.5	71.5	71.0	73.5	75.0	72.0	72.5	71.5	71.5	74.0	74.5
		Mean	79.9	79.0	78.0	79.4	80.8	82.1	79.5	79.9	80.5	80.4	81.7	81.3
		75%	83.5	83.0	83.0	83.0	90.5	90.5	88.5	88.5	89.5	90.5	90.5	90.5
		Max.	88.0	90.0	89.0	89.0	93.0	97.0	92.0	92.0	93.0	96.0	95.0	98.0
		Median	79	78	76	79	79	81	79	80	78	79	81	82
		Std.	8.11	8.09	7.70	7.67	9.03	8.83	8.87	8.73	9.89	9.71	9.18	9.30
	Significance of difference	N-N	o	o	o	–	o	o	o	–	o	o	o	–
Skin temperature (°C)	Statistic result	Min.	32.7	32.3	31.6	32.1	31.5	31.5	30.9	30.8	30.1	30.5	30.0	30.2

(continued)

Table 4.5 (continued)

Categories			26 °C				21 °C				18 °C			
			S-N	S-N-1m	S-N-1y	N-N	S-N	S-N-1m	S-N-1y	N-N	S-N	S-N-1m	S-N-1y	N-N
		25%	33.5	32.8	33.1	33.1	32.6	31.6	31.2	31.2	31.8	31.6	31.2	30.9
		Mean	34.1	34.0	33.9	33.9	32.9	32.8	32.4	32.3	32.2	32.1	31.8	31.7
		75%	34.8	35.0	34.9	34.8	33.3	33.5	33.1	33.1	32.9	32.5	32.3	32.1
		Max.	35.4	35.7	35.3	35.5	34.5	34.6	34.2	34.1	33.9	33.8	33.5	33.4
		Median	34.1	34.6	34.6	34.4	32.7	32.7	32.4	32.2	32.2	32.3	31.9	31.7
		Std.	0.95	1.31	1.37	1.22	0.98	1.17	1.22	1.21	1.26	1.09	1.16	1.09
	Significance of difference	N-N	o	o	o	–	*	o	o	–	*	o	o	–
Normalized skin blood flow	Statistic result	Min.	–	–	–	–	0.89	0.84	0.77	0.70	0.67	0.64	0.66	0.66
		25%	–	–	–	–	0.91	0.87	0.84	0.81	0.75	0.72	0.70	0.68
		Mean	1.00	1.00	1.00	1.00	0.93	0.89	0.86	0.82	0.80	0.76	0.74	0.71
		75%	–	–	–	–	0.95	0.90	0.88	0.82	0.86	0.80	0.79	0.72
		Max.	–	–	–	–	1.24	0.95	1.13	0.98	0.97	0.88	0.86	0.73
		Median	–	–	–	–	0.93	0.89	0.86	0.82	0.80	0.74	0.74	0.70
		Std.	–	–	–	–	0.08	0.03	0.08	0.05	0.08	0.06	0.06	0.02
	Significance of difference	N-N	–	–	–	–	**	**	*	–	**	**	**	–

heart rate as well. When the temperature dropped from 26 to 18 °C, the heart rate increased slightly.

For skin temperature and skin blood flow, there did exist significant differences between N-N group and the other three groups. Especially between-group N-N and group S-N, the p values of the T-test were smaller than 0.05. But the difference between group N-N and group S-N-1y was less obvious, even exhibiting no difference in some cases. Additionally, both skin temperature and skin blood flow can be affected by thermal conditions. When the temperature dropped from 26 to 18 °C, the mean skin temperature decreased from 33.9 to 32.0 °C. Meanwhile, the average blood flow dropped from 347 to 262 PU, a decrease of nearly 25%.

It should be noted that the perfusion unit (PU) normally equals to measured blood cell concentration multiplied by the average velocity of blood cells in a specific area. However, since the absolute blood cell concentration and absolute velocity of blood cells are not known, it is meaningless to compare the absolute PU values. To address this issue, the skin blood flow results were expressed as a percentage of the baseline (each individual's PU value in the neutral temperature case).

Figure 4.12 compares the four groups' thermal sensation and comfort perceptions. The independent T-test results are labeled as '*' when $p < 0.05$, '**' when $p < 0.01$, and are not labeled when $p > 0.05$. Compared with N-N group, the other three groups felt less cold and less uncomfortable when they were exposed to 21 and 18 °C. But in the neutral temperature case (26 °C), each group had similar subjective perceptions. This phenomenon is consistent with the physiological data shown in Table 4.5. Figure 4.12 also shows the calculated shivering rate, based on the subjects' self-reported shiver. Generally, N-N subjects had higher shivering rate than other groups. Even among the groups who migrated from cold indoors, their shivering rate increased with the longer time spent in warm-neutral indoors.

4.4.3 The Time-Scale of Physiological Acclimation

To show how does time work on human thermal adaptation, Fig. 4.13 illustrates the dynamic process of physiological adaptation (on the basis of skin blood flow and skin temperature) and subjective adaptation (on the basis of average TSV and TCV). It can be seen that migrating subjects' physiological response and subjective perception converged gradually toward the same values with group N-N over an adaptation time. The longer the time for adaptation, the closer the group becomes to the N-N subjects. However, the 'gap' between N-N subjects and S-N subjects didn't disappear completely even after 1-year's adaptation, especially for the two physiological indexes. It suggests that full human thermal adaptation may require considerable time.

Although there is only a small body of research focusing on human thermal adaptation process, some clues relevant to the time-scale of human thermal adaptation can be gleaned. Table 4.6 summarizes some descriptions and main results of these

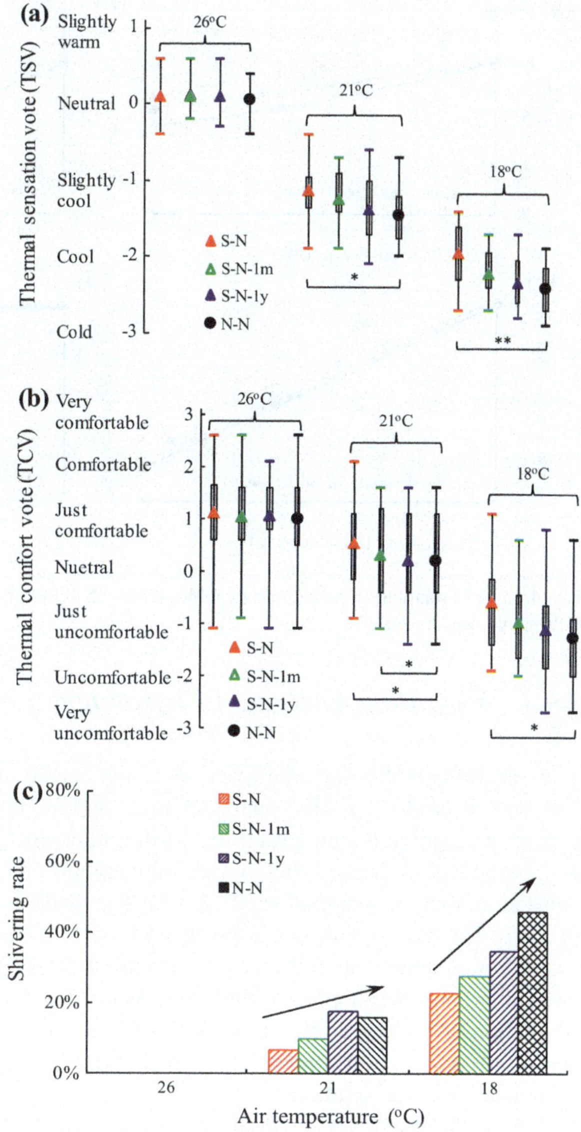

Fig. 4.12 Subjective perceptions: **a** thermal sensation vote, **b** thermal comfort vote, **c** shivering rate

studies. It is interesting to see that the reported adaptation time varied from days to years. For those who migrated to new thermal environments, the adaptation process may start from a few days or weeks after their move but will require months or years to be accomplished completely. These seemingly chaotic results once again indicate that human thermal adaptation requires quite a long time instead of being accomplished in one stroke.

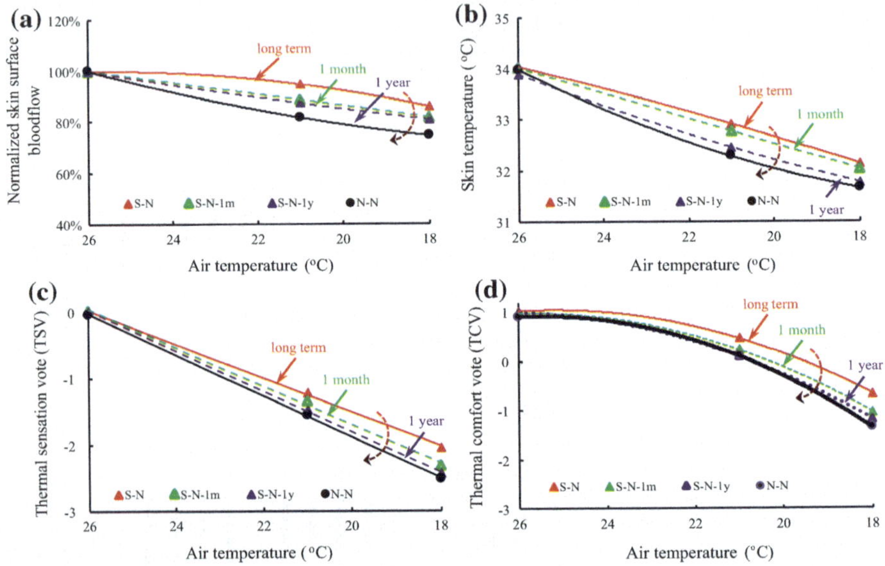

Fig. 4.13 The adaptation process: **a** skin blood flow; **b** skin temperature; **c** thermal sensation vote; **d** thermal comfort vote

4.4.4 Weighting Factors of Physiological Acclimation

The variability in thermal adaptive time-scales can be caused by different reasons, but one reason should be attributed to different adapting layers. It is believed that human thermal adaptation is an interaction of different layers like physiological acclimation, psychological adaptation, behavioral adjustment, and culture adaptation. The time constant of these layers may vary from days or weeks for behavior changes to seasons or years for the physiological layer, and even to decades or lifetime for culture adaptation. Coupled with the cross impacts of these layers, it is rather complicated to acquire an exact adaptation time. But, which layer is the driver of human thermal adaptation? And how about the weighting factor of these adaptation approaches? To answer these questions, a new index is proposed to determine the dynamic process of physiological adaptation.

Firstly, three indexes are defined: $BD_{group\ i}$, the basic difference between group i and group N-N in a neutral condition (26 °C); LD, the long-term difference between group S-N and group N-N; GD_i, the group difference between group i and group N-N in different time-scales. Specific definitions of these indexes were expressed as follows:

$$BD_{group\,i} = \text{Ave.}_{26.\text{group i}} - \text{Ave.}_{26.\text{group N-N}} \tag{4.4}$$

$$\begin{aligned} GD_i = {} & (\text{Ave.}_{21.\text{group i}} - \text{Ave.}_{21.\text{group N-N}}) \\ & + (\text{Ave.}_{18.\text{group i}} - \text{Ave.}_{18.\text{group N-N}}) - 2 * BD_{group\,i} \end{aligned} \tag{4.5}$$

Table 4.6 Comparison of adaptation time-scales

Literature	Method	Parameters investigated	Adaptation description	Experimental condition	Adapting time
van der Lans et al. [8]	Chamber experiment	Brown adipose tissue, Energy expenditure, TSV, TCV, etc.	10 consecutive days cold protocol	Cold exposure	10 consecutive days
Cao et al. [13]	Chamber experiment	Rectal temperature, skin temperature, sweat rate	From tropical area to temperate area	Hot exposure	Nearly 4 years
Yu et al. [14]	Chamber experiment	Heat stress protein, skin temperature	From naturally ventilated environment to air-conditioned environment	Hot exposure	More than 2 years
Schweiker and Wagner [1]	Field study	TSV, TCV, Clothing insulation	From cold indoors to neutral-warm indoors	–	A heating season (nearly 5 months)
Present study	Chamber experiment	Skin blood flow, skin temperature, TSV, TCV, etc.	From cold indoors to neutral-warm indoors	Cold exposure	From 1 month to over 1 year

$$\begin{aligned}\mathrm{LD} &= (\mathrm{Ave.}_{21.\mathrm{group\,S-N}} - \mathrm{Ave.}_{21.\mathrm{group\,N-N}}) \\ &\quad + (\mathrm{Ave.}_{18.\mathrm{group\,S-N}} - \mathrm{Ave.}_{18.\mathrm{group\,N-N}}) - 2 * BD_{S-N}\end{aligned} \tag{4.6}$$

Then, LD is regarded as a baseline, which is referred to as 100%. The adaptation rate is calculated according to the following equation:

$$\text{Adaptation rate} = \frac{GD_i}{LD} \tag{4.7}$$

Figure 4.14 illustrates the calculated adaptation rate. It reveals the dynamic adapting process of human thermal adaptation to some extent. The difference between migrated groups and the N-N group narrowed gradually as time proceeds. For subjective perceptions like TSV and TCV, the migrated subjects become almost the same as the N-N group after one year's adaptation. But for physiological response like skin

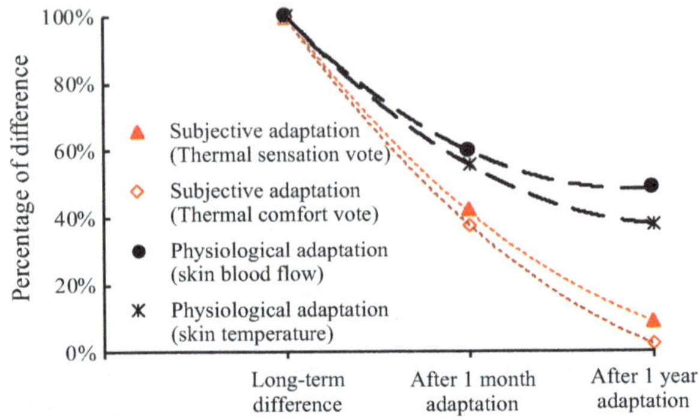

Fig. 4.14 The adapting rate of human thermal adaptation

blood flow and skin temperature, there still existed more than 40% difference even after a whole year of adaptation. This phenomenon suggests that the physiological adaptation process lags behind the overall subjective adaptation.

4.5 Short-Term Physiological Regulation

The above two sections found that the metabolic rate of human body is significantly affected by temperature conditions. Given the metabolic rate can directly reflect the heat production of human body, it is very important for human body heat balance model and thermal perception.

To illustrate the importance of metabolic rate in thermal comfort prediction, Fig. 4.15 presents the impacts of the metabolic rate change on PMV variance at two clothing insulation levels. When metabolic rate increases from 0.9 to 1.5 met (1 met equals to 58.2 W/m^2), it can result in more than 2 °C variance in the predicted neutral temperature (temperatures when PMV equals to 0) or approximately 1.5-unit

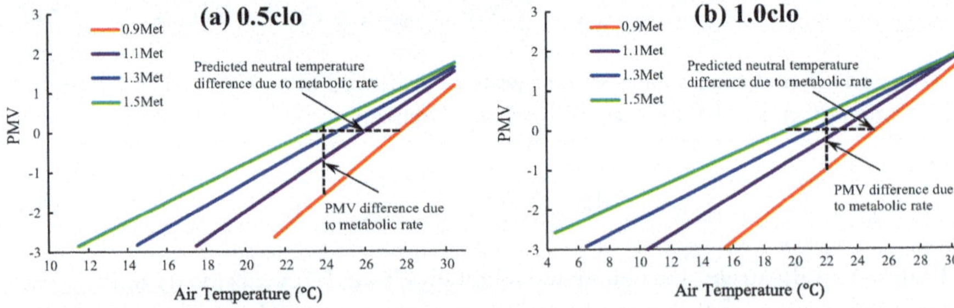

Fig. 4.15 Effects of metabolic rate on PMV variance at two clothing insulation levels

scale of PMV difference. Apart from PMV calculation, other thermal comfort models also need metabolic rate as an essential prerequisite for human body heat balance calculation [15]. Not surprisingly, human metabolism has been treated as a basic and crucial index for thermal comfort prediction.

4.5.1 Experiment Supplement

Based on the previous two sections, this section focused more on metabolic rate changes in different temperature conditions. The following supplementary experiments were carried out.

(1) To choose reasonable experimental conditions, PMV of each temperature case was calculated in advance basing on the assumptions that *M* is 1.1 met, *v* is 0.1 m/s, *RH* is 50%. As shown in Fig. 4.16, experimental conditions with PMV values ranging from 1.84 (warm) to −3.94 (intolerable cold) were selected. Temperatures ranging from neutral (26 °C) to cold (24/21/18/16 °C) and to hot (31/28 °C) were created in a climate chamber. In the preparing phase, subjects had 10 min to calm down, change clothes, learn how to vote, and to wear physiological sensors. To create diverse thermal conditions, two sets of clothes with different insulation values were utilized, namely: light clothing set (underwear, T-shirt, shorts, socks, and sport shoes) with a total clothing insulation of 0.42 clo, and medium clothing set (underwear, T-shirt, sweater, trousers, socks, and sport shoes) with a total clothing insulation of 0.91 clo. In the second phase, all the subjects stayed in a neutral temperature (26 °C) for 30 min. Then, they moved

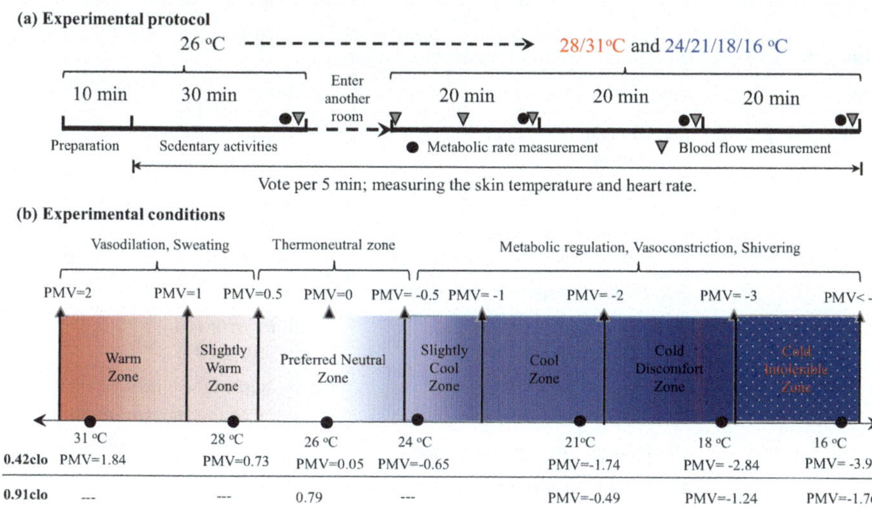

Fig. 4.16 Experimental design

to other cold conditions (24/21/18/16 °C) or warm conditions (31/28 °C) for another 60 min. During the experimental period, all the participants remained seated and engaged only in sedentary activities (Table 4.7).

(2) According to ISO 8996 [16], there are four levels of accuracy and eight methods to measure human body metabolic rate (as listed in Table 4.8). The more accurate the method, the more expensive and time-consuming it is. For instance, method 4B estimates the average metabolic rate over a long period of 1–2 weeks. Method 4C needs the elaborate chamber to measure body heat loss. Considering the purpose of this study, we utilized method 4A, namely estimating metabolic rate through oxygen and carbon dioxide measurement.

Table 4.9 analyses the impact of such errors on metabolic rate calculation method 4A. The measurement of height and weight can be easily conducted in laboratory settings. At the same time, an error of ±1 cm in height or an error of ±0.1 kg in weight can only result in <1% error in metabolic rate. As for Q_{CO_2} and Q_{O_2}, their

Table 4.7 Subjects' profile in the third phase test

Group	Clothing insulation	Sample sizes				Age (year)	Height (m)	Weight (kg) 28, 31 °C
		28, 31 °C	26 °C	24 °C	21, 18, 16 °C			
Cold side	0.42 clo	–	30	30	30	20.5 ± 1.5	0.42 clo	–
	0.91 clo	–	(15)	–	15	20.1 ± 1.3	0.91 clo	–
Warm side	0.42 clo	15	(15)	–	–	19.2 ± 1.8	0.42 clo	15

Table 4.8 Levels for the determination of the metabolic rate

Level	Method	Accuracy
1. Screening	1A: Classification according to occupation	Rough information Very great risk of error
	1B: Classification according to activity	
2. Observation	2A: Group assessment tables	High error risk Accuracy: ±20%
	2B: Tables for specific activities	
3. Analysis	Heart rate measurement under defined conditions	Medium error risk Accuracy: ±10%
4. Expertise	4A: Measurement of oxygen consumption	Errors within the limits of the accuracy of the measurement or of the time and motion study Accuracy: ±5%
	4B: Doubly labeled water method	
	4C: Direct calorimetry	

Table 4.9 Impact of measuring error on the metabolic rate

Measuring	Possible error	Error analysis	Possible error in *M* (%)
$\frac{\Delta Q_{CO_2}}{Q_{CO_2}}$ (L/min)	±2%	$< \left\lvert \frac{\Delta Q_{CO_2}}{Q_{CO_2}} \right\rvert \cdot 100\%$	<2
$\frac{\Delta Q_{O_2}}{Q_{O_2}}$ (L/min)	±2%	$< \left\lvert \frac{\Delta Q_{O_2}}{Q_{O_2}} \right\rvert \cdot 100\%$	<2
H (m)	1 cm	$0.725 \cdot \left\lvert \frac{\Delta H}{H} \right\rvert \cdot 100\%$	<0.5
W (kg)	0.1 kg	$0.425 \cdot \left\lvert \frac{\Delta W}{W} \right\rvert \cdot 100\%$	<0.5

measuring method is much more difficult. And the error in these two entities has a greater impact on metabolic rate calculation. For instance, a 2% error in Q_{CO_2} will lead to 2% error in the estimation of metabolic rate.

It should be noted that the purpose of the present study is to explore the impacts of indoor thermal conditions on occupants' metabolic heat production. Therefore, we mainly focus on the variation trend of metabolic rate with thermal conditions, rather than the accuracy of the absolute value of metabolic rate. Although there exist measuring errors in metabolic rate determination, the general changing is reliable.

4.5.2 *Metabolic Rate Changes and Its Description*

Figure 4.17 illustrates the metabolic rate in different thermal conditions. Despite the individual differences, the overall metabolic rates increase when thermal conditions depart from the neutral zone. Especially in cold conditions, the increasing trend is more obvious. For instance, the metabolic rate in 16 °C with 0.42 clo is 15.8%

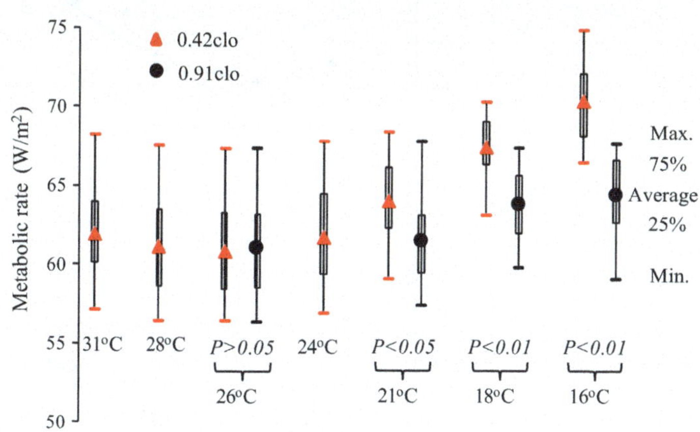

Fig. 4.17 Distribution of metabolic rate in different temperature and clothing cases (seated, sedentary)

higher than that in 26 °C. Apart from temperature, clothing insulation also matters. In cold conditions, the medium clothing set situation (0.91 clo) greatly mitigated the increasing amplitude of metabolism heat production. The average metabolic rate in 16 °C with 0.91 clo is 9.2% lower than that with 0.42 clo at the same temperature.

It should be noted that besides the metabolic rate, other physiological indexes are also affected by thermal condition changes. Given that the purpose of this section is to establish an adaptive heat balance prediction model, we only introduce the metabolic rate changes which can directly reflect human body heat production.

The above results collectively point to the fact that human thermoregulation can be significantly affected by thermal conditions such as ambient temperature and clothing insulation. However, the thermal condition is a synthesis of many factors like temperature, relative humidity, air velocity, clothing levels and etc. It is not easy to examine the individual impact of all these factors through chamber experiment. Therefore, we chose TSV and PMV as an independent variable to describe the changing rate of metabolic heat production. Generally, the variation trend of metabolic rate can be well expressed by a second-order polynomial equation. A neutral thermal sensation corresponds to the lowest metabolic value while 'cold' or 'hot' sensations correspond to higher metabolic heat production. Considering the absolute values of the metabolic rate may vary with different activity levels, Fig. 4.18 presents the modified equations for relative changes in metabolic rate.

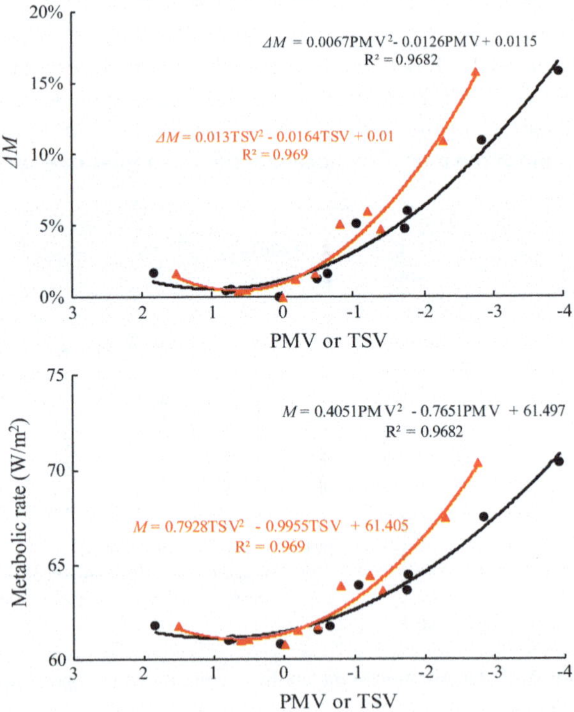

Fig. 4.18 Average metabolic rate as a function of experimental temperature (seated, sedentary)

The modified equations are supposed to serve as references and be applied in both thermal comfort field studies and modeling simulation. The equation with TSV is suitable for thermal comfort studies (Eq. 4.8). Because TSV is available in most field studies with subjective questionnaires. Meanwhile, the equation with PMV can be accepted in human body heat balance calculation models (Eq. 4.9). Because PMV itself is an index based on the heat balance equation of human body.

$$\Delta \mathrm{M} = 0.013\mathrm{TSV2} - 0.0164\mathrm{TSV} + 0.01 \tag{4.8}$$

$$\Delta \mathrm{M} = 0.0067\mathrm{PMV2} - 0.0126\mathrm{PMV} + 0.0115 \tag{4.9}$$

4.6 Conclusions

This chapter examined how physiological acclimation affects thermal comfort through three comparative climate chamber experiments. The main conclusions are as follows.

(1) Physiological acclimatization can be reflected in the regulations of several physiological parameters when responding to short-term thermal stress and long-term thermal adaptation. For example, our body can adapt itself to short-term thermal conditions changes through metabolic rate and skin temperature changes. At the same time, the long-term thermal experience can also affect our body's response to thermal stimulus through differences in physiological indicators such as skin temperature and skin blood flow.

(2) The long-term physiological acclimation can be described by the "weighting factor" and "adaptive coefficient" proposed in this chapter, in which the weighting factor can reflect the proportion of physiological acclimation to overall thermal adaptation while the adaptation coefficient describes how physiological acclimation changes with time. It was found that 40% overall thermal adaptation was on account of long-term physiological acclimatization. Also, physiological acclimatization lags behind the subjective adaptation.

(3) The short-term physiological regulation can be reflected by metabolic rate changes. It was found that human body metabolism can be influenced by ambient thermal conditions (such as temperature and clothing insulation). For example, the metabolic rate will be higher in both cold and hot conditions, and the changing trend can be described by equations with independent indexes of PMV or TSV.

References

1. Schweiker M, Wagner A (2015) A framework for an adaptive thermal heat balance model (ATHB). Build Environ 94:252–262
2. de Dear R, Akimoto T, Arens E, Brager G et al (2013) Progress in thermal comfort research over the last twenty years. Indoor Air 23(6):442–461
3. van Marken Lichtenbelt W, Kingma B, van der lans A, Schellen L (2014) Cold exposure-an approach to increasing energy expenditure in humans. Trends Endocrinol Metab 25(4):165–167
4. Zhu Y, Ouyang Q, Cao B, Zhou X, Yu J (2016) Dynamic thermal environment and thermal comfort. Indoor Air 26(1):125–147
5. Johnson F, Mavrogiann A, Ucci M et al (2011) Could increased time spent in a thermal comfort zone contribute to population increases in obesity. Obes Rev 12(7):543–551
6. Kingma B, Frijns A, Schellen L et al (2014) Beyond the classic thermoneutral zone including thermal comfort. Temperature 1(2):142–149
7. Yu J, Ouyang Q, Zhu Y et al (2012) A comparison of the thermal adaptability of people accustomed to air conditioned environments and naturally ventilated environments. Indoor Air 22:110–118
8. van der Lans A, Hoeks J, Brans B et al (2013) Cold acclimation recruits human brown fat and increases no-shivering thermogenesis. J Clin Invest 123(8):3395–3403
9. Ramanathan NL (1964) A new weighting system for mean surface temperature of the human body. J Appl Physiol 19(3):531–533
10. Nishi Y (1981) Measurement of thermal balance in man. Stud Environ Sci 10:29–39
11. van Marken Lichtenbelt W, Vanhommerig J, Smelders N et al (2009) Cold activated brown adipose tissue in healthy men. N Engl J Med 360(15):1500–1508
12. Hanssen M, Hoeks J, Brans B, van der Lans A et al (2015) Short-term cold acclimation improves insulin sensitivity in patients with type 2 diabetes mellitus. Nat Med 21(8):863–865
13. Cao B, Zhu Y, Ouyang Q, Zhou X, Huang L (2011) Field study on human thermal comfort and thermal adaptability during the summer and winter in Beijing. Energy Build 43(5):1051–1056
14. Yu J, Cao G, Cui W, Ouyang Q, Zhu Y (2013) People who live in a cold climate: thermal adaptation differences based on availability of heating. Indoor Air 23(4):303–310
15. Ala-Juusela M, Shukuya M (2014) Human body exergy consumption and thermal comfort of an office worker in typical and extreme weather conditions in Finland. Energy Build 76:249–257
16. Luo M, Ji W, Cao B et al (2016) Indoor climate and thermal physiological adaptations: evidences from migrants with different cold indoor exposures. Build Environ 98:30–38

Chapter 5
Personal Control and Its Phycological Effects on Thermal Adaptation

5.1 Background

In previous chapters, we discussed the physiological acclimation effects on building occupants' thermal adaptation. However, psychological adaptation, as an important approach for thermal adaptation, has great controversy, so further exploration is needed.

5.1.1 Personal Control of Indoor Thermal Environment

Systems and devices that heat or cool individual occupants (or small groups of occupants) have widely existed in buildings. Driven by the benefits of both comfort improving and energy-saving [1, 2], various forms of desk, wall, and ceiling fans, radiant or convective heaters, and temperature-controlled surfaces on chairs, desks, and floors, have been available in the marketplace. These devices are feasible to meet individual comfort requirements which differ due to variation in gender, age, body mass, metabolic rate, clothing, and thermal adaptation [3–5]. So that they can help to achieve a much higher satisfaction rate than it is in uniformly conditioned space [6]. Brager et al. [7] conducted a field study in a naturally ventilated office building, which showed occupants with more opportunities to operate windows voted thermal sensation closer to neutral than those who had less capability to control windows. Langevin et al. [8] re-analyzed the data of three case buildings from the RP-884 database in a more detailed way and reported significant correlations indeed appeared between comfort evaluation and perceived control variables even in air-conditioned buildings. The concept and benefit of personal environmental control have been well summarized recently in Ref. [9] and Fig. 5.1.

M. Luo, *The Dynamics and Mechanism of Human Thermal Adaptation in Building Environment*, Springer Theses,
https://doi.org/10.1007/978-981-15-1165-3_5

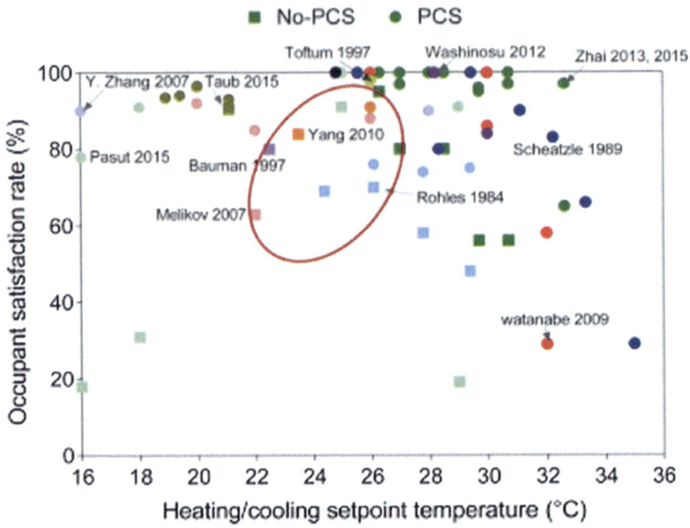

Fig. 5.1 Higher comfort rates achieved at warmer- or cooler-than-neutral ambient conditions (neutral conditions shown as squares in ellipse) [9]

5.1.2 *The Missing Linkage Between Personal Control and Thermal Comfort*

All the above studies provide some references related to positive effects of personal control on indoor thermal environment, while there also exists difference voices saying that personal control had no impact on occupants' comfort or working performance [10]. Based on the current small body of empirical evidence, researchers haven't reached a consensus on whether and how personal control affects thermal comfort perception yet. When personal control approaches were utilized by occupants, the physical thermal conditions would be changed. We need to figure out whether the observed comfort improvement is due to psychological effects, or it is because of physical condition changes. Zhou et al. [11] once demonstrated that personal comfort approach can improve thermal comfort perception through pure psychological effects. However, the previous studies didn't thoroughly clarify the linkage between personal control capability and thermal comfort. So that it is unable to incorporate its psychological effects into the thermal adaptation model.

5.1.3 *Objectives of This Chapter*

This study aims to figure out: (1) Whether personal control on indoor thermal environment has positive effects on thermal comfort and how much it could be. (2) Whether the effect is purely caused by psychological effects or other physical changes? (3) Is it possible to describe the psychological effects in a straightforward way? Note, the psychological effect of personal control on space conditioning systems was referred to as "***Perceived Control***".

5.2 Personal Control and Thermal Comfort Improvement

5.2.1 Field Investigation Design

As the reflection of the influence of personal control on indoor thermal environment, the disputing issue on centralized and decentralized building heating systems is a typical example. Comparing these two heating forms is of practical significance as well as helping to validate the effects of personal control. Therefore, we carried out a traditional field thermal comfort investigation to compare occupants' thermal comfort perceptions in these two heating forms.

As shown in Table 5.1, occupants in apartments with district heating supply (BJ-A) had no direct way to control the indoor thermal environment, meaning this group of occupants had to accept their thermal environment passively. For group BJ-B, occupants could control the indoor thermal environment by adjusting individual space heating systems. While other aspects of the selected apartments, such as the presence of operable windows, the age of the building, and the spatial layout, were similar in the initial background survey. As all the buildings sample were located in the same community, no significant differences in other factors like age, gender, and annual income.

The investigation included simultaneous measurements of physical parameters and subjective questionnaires. Table 5.2 lists the information related to these two complementary levels of investigation. The detailed level aimed to record the indoor thermal conditions and occupants perceived thermal comfort in minute detail, while

Table 5.1 Summary of the surveyed buildings

Groups	Location	Survey period	Number of apartments	Space heating system	Personal control over indoor thermal environment
BJ-A	Beijing	Dec. 2012–Feb. 2013	Background level: 45 Detailed level: 15	District heating supply	No
BJ-B	Beijing	Dec. 2012–Feb. 2013	Background level: 41 Detailed level: 17	Individual household gas boiler heating	Free to adjust

Table 5.2 Content of measurements and questionnaires

	Physical measurements	Survey questionnaires	
		Building information	Comfort performance
Background study	Ta, Tg, and RH	Building characteristics, lifestyle, etc.	Clothing situation, TSV, TCV, thermal preference, thermal acceptance, etc.
Detailed study	Ta, Tg, ν, RH, and meteorological station		

the goal of the background level was to extend the scope of the research and verify the results of the detailed level investigation. The apartments in the detailed study were observed continuously, and those in the background investigation were examined only once.

Indoor thermal parameters were continuously measured at 5-minute intervals with different instruments. In each surveyed apartment, sensors with self-recording loggers were placed about 1.0 m above the floor in the living room, bedroom, and reading room. Occupants' subjective sensations were determined by questionnaires measuring thermal sensation vote (TSV), thermal comfort vote (TCV), thermal preferences, thermal acceptance, etc. The TSV adopted a seven-point scale ranging from cold (−3) to hot (+3), with neutral (0) in the middle [12]. Thermal preferences were assessed using a three-point scale: 'warmer', 'maintained', and 'cooler'. Clothing insulation was calculated in accordance with ISO 9920-2009 [13]. In total, 900 valid questionnaires and corresponding indoor thermal conditions were collected.

5.2.2 *Indoor Thermal Environment Comparison*

Table 5.3 and Fig. 5.2 shows the measured indoor thermal parameters corresponding to the questionnaires. Both BJ-A and BJ-B apartments had indoor thermal environment met the requirements of heating design code with temperature ranging from 18 to 24 °C and relative humidity ranging from 20 to 50%. At the same time, both groups had overheated or too cold room temperatures.

However, it is worth noting that for group BG-A residents, the heating terminals in their apartments were not adjustable and they couldn't actively control the room temperature even when there occurred 27.4 °C high temperature. For residents in group BJ-B, their room temperatures could be freely controlled. Although some cold temperatures like 14.5 °C appeared, it was the result of the users' active selection.

Table 5.3 Statistical summary of indoor climatic and thermal comfort indices

Parameter	Group BJ-A				Group BJ-B			
	Mean	Std	Min	Max	Mean	Std	Min	Max
CLO	0.86	0.20	0.35	1.48	0.87	0.22	0.41	1.51
MET	1.2	0.1	1.0	1.4	1.2	0.1	1.0	1.5
Ta (oC)	21.3	2.8	14.4	28.4	21.2	2.6	14.5	25.2
MRT (oC)	21.7	2.6	14.8	28.4	21.3	2.0	14.3	25.5
Top (oC)	21.5	2.8	14.6	28.4	21.3	2.7	14.3	25.3
ν (m/s)	0.02	0.04	0.00	0.56	0.02	0.04	0.00	0.28
RH (%)	34.4	9.7	11.7	61.5	36.9	10.0	10.6	68
PMV	−0.4	0.5	−1.7	0.9	−0.6	0.4	−1.9	0.5

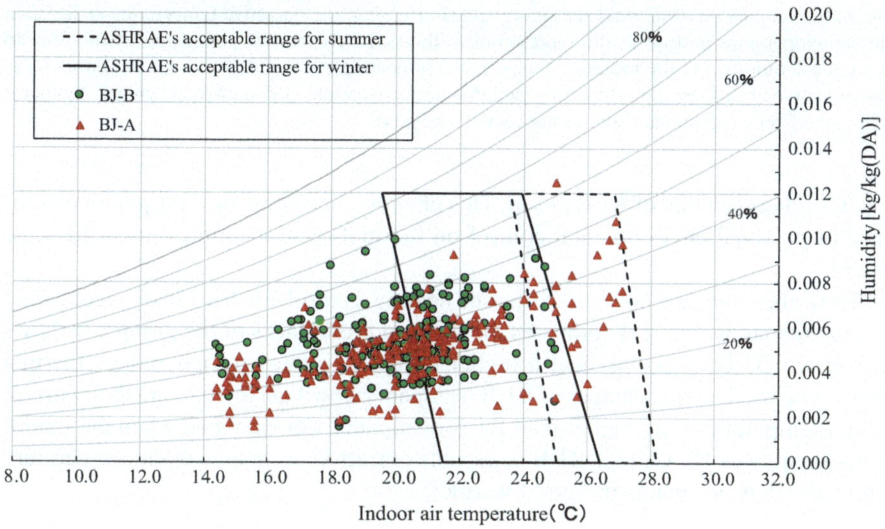

Fig. 5.2 Indoor thermal environment distribution

5.2.3 *Thermal Comfort Comparison*

In order to analyze the influence of environmental control capability on thermal comfort, Fig. 5.3 compares the actual thermal sensation voting (TSV) between BJ-A and BJ-A groups. It shows that, when the room temperature was relatively cold (14–20 °C), the BJ-B residents felt warmer than the BJ-A group. The thermal sensation of BJ-B group changed more smoothly with temperature and was basically near thermal neutrality (−0.5 < TSV < 0.5). In contrast, the TSVs from BJ-A group were more dispersed, with a thermal neutral temperature (TSV = 0) of 20.9 °C which is

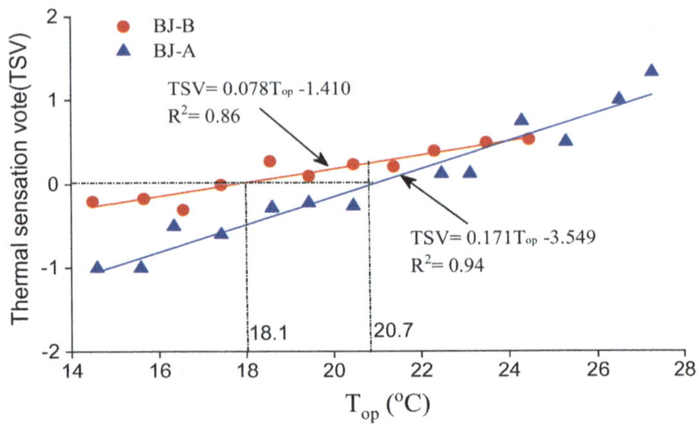

Fig. 5.3 Comparison of thermal sensation vote (In Fig. 5.4, the operative temperature rather than the air temperature is used as the representative thermal environment parameter, mainly due to two considerations: (1) the radiation temperature in heating environment cannot be ignored; (2) The objective of this section is to verify the influence of personal environmental control. Neither it affects the later experimental studies nor model analysis)

2.8 °C higher than that of BJ-B group. This phenomenon confirmed the positive influence of personal environmental control on thermal comfort improvement to some extent.

To further analyze how did different heating control modes satisfy occupants' heating demands, Fig. 5.4 compares the thermal preferences of the groups. It shows that group BJ-B occupants had a higher acceptance rate on their indoor thermal environments. The percentages of BJ-B occupants who want to maintain their current indoor temperatures are higher than BJ-A occupants. For example, when the indoor temperature was 15 °C, 18% BJ-B occupants preferred no change in the temperature while all BJ-A occupants preferred warmer.

5.2.4 *Adaptive Behaviors*

To illustrate how occupants adapt themselves to the environment with and without personal control over indoor thermal conditions, Fig. 5.5 shows the adaptive behaviors among groups BJ-A and BJ-B. There are many forms of adaption in residential buildings, such as adjusting clothing, opening windows, drinking cold or warm drinks, etc. However, occupants in BJ-B apartments have more effective approaches for dealing with non-neutral thermal conditions because they can create preferable indoor thermal conditions by setting the space heating temperature lower or higher as necessary. This may be a plausible explanation for why BJ-B apartments have a higher percentage of occupants who want to maintain their indoor thermal conditions. Upon closer examination of the frequency of each adjustment approach in

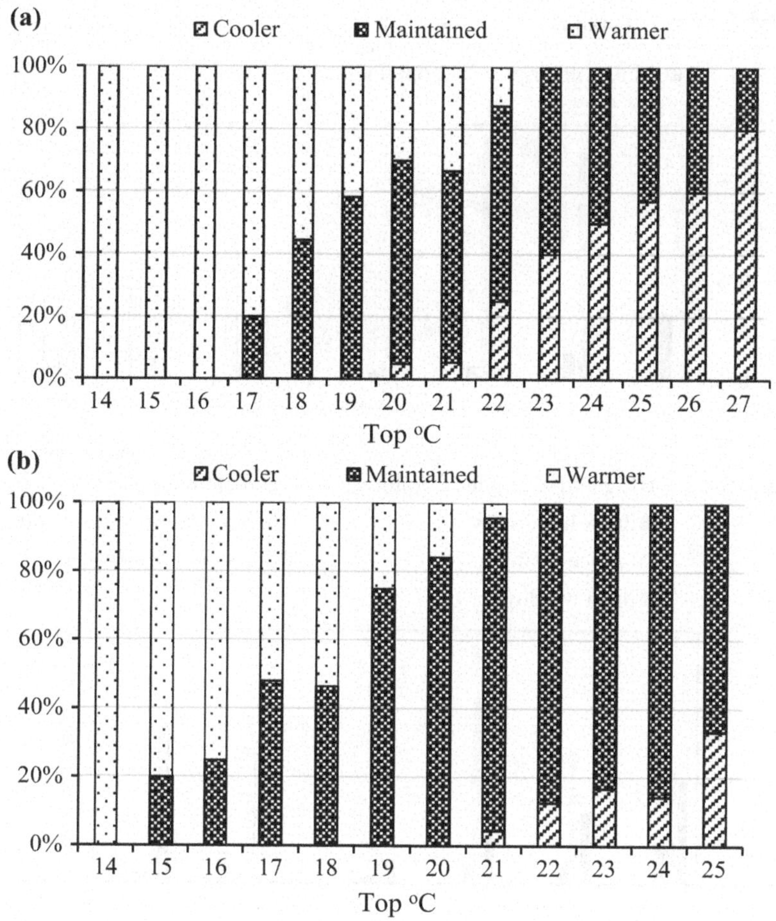

Fig. 5.4 Personal control and thermal expectations: **a** BJ-A; **b** BJ-B

BJ-B buildings, it is interesting that occupants tended to choose a lower heating temperature first when they felt hot but chose to add clothing first whey they felt cold. We attribute this phenomenon to economic effects, as a lower space heating temperature will result in lower heating fees.

In order to analyze the influence of environmental control mode on heating behavior, Fig. 5.6 shows three room temperature profiles from BJ-B group. Among which, apartment A and B had a similar averaged room temperature around 23.5 °C while that of apartment C is 19.2 °C, 4° lower. As for fluctuation, A and C had similar temperature fluctuation patter while B had a stable temperature profile. The results show that the heating energy consumption of A and B were very close, both around about 10 $m^3/(m^2\ a)$. Apartment C consumed much less natural gas for heating, only 8.5 $m^3/(m^2\ a)$.

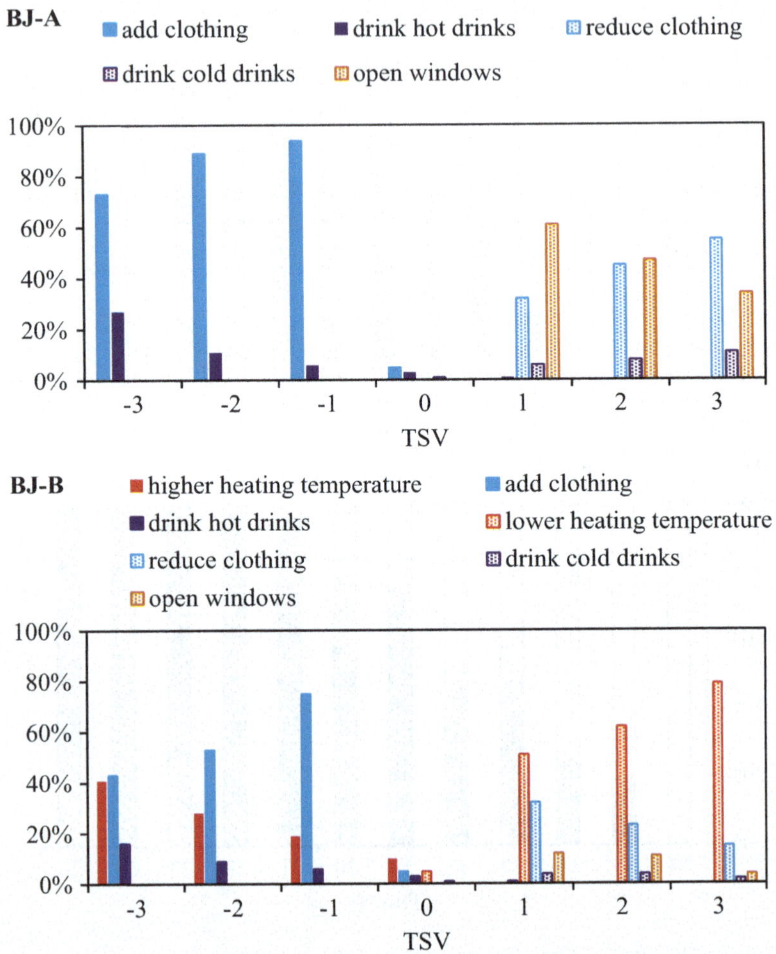

Fig. 5.5 Adaptive behaviors

Figure 5.7 shows the influencing factors of household heating gas consumption. It indicates that the heating gas consumption mainly depends on the heating mode (i.e. the heating room temperature and the heating time). The floor location also has a certain influence. For example, the top-level apartments tended to consume more natural gas for heating purpose.

Figure 5.8 shows the heating gas consumption during the investigation. Among which, the average of background survey is 8.4 $m^3/(m^2 \ a)$ while that of the detailed survey is 8.2 $m^3/(m^2 \ a)$. Based on the survey, Table 5.4 compares the cost of two heating control systems. If only the gas consumption is considered, the cost for individually controlled heating ranges from 1154 (RMB per year) to 2554 (RMB per year). Even with the initial investment, it still has economy advantage. Therefore,

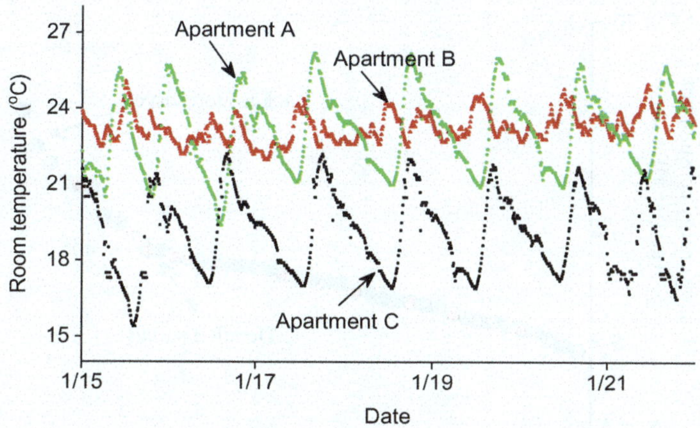

Fig. 5.6 Room temperature profiles in BJ-B apartment

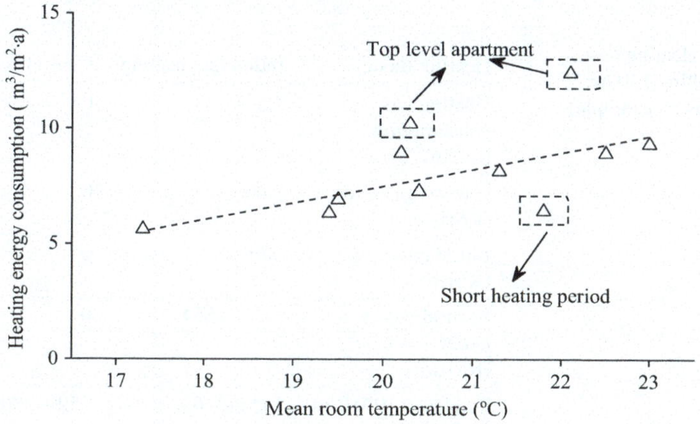

Fig. 5.7 Factors affecting space heating energy consumption

compared with the centralized heating system with uncontrollable room temperature, individual household heating gives users the option to heat according to actual demands, thus balancing comfort and economy.

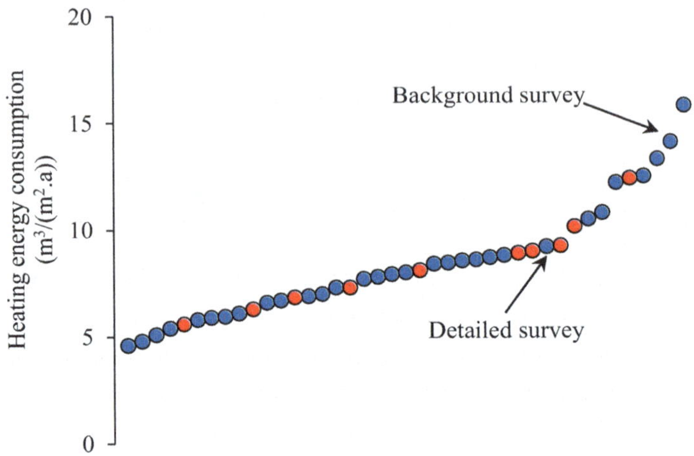

Fig. 5.8 Heating energy consumption for different apartments

Table 5.4 Heating cost analysis (taking 100 m^2 apartment as an example)

Heating mode	Individual heating	Centralized heating
Heating gas consumption ($m^3/(m^2$ a))	5.6–12.4	0
Natural gas price (RMB/m^3)	2.06	0
Initial cost (RMB/a)	600	0
Natural gas expenditure (RMB/a)	1154–2554	0
Total cost (RMB/a)	1754–3163	2500–3000

5.3 The Missing Linkage Between Personal Control and Thermal Comfort

5.3.1 Experiment Design

The overall experimental protocol is shown in Fig. 5.9. Considering the fact that air temperature has an obvious effect on comfort perception and is also easy to control, it was chosen as the control variable in this experiment. And five temperature cases (31, 28, 26, 24, 21 °C), ranging from hot to cold conditions, were selected to represent different thermal environments. During the experiment, air temperatures (five temperature cases were ordered randomly), relative humidity (maintained in 50% ± 5%) and globe temperatures (similar with air temperatures) were monitored by

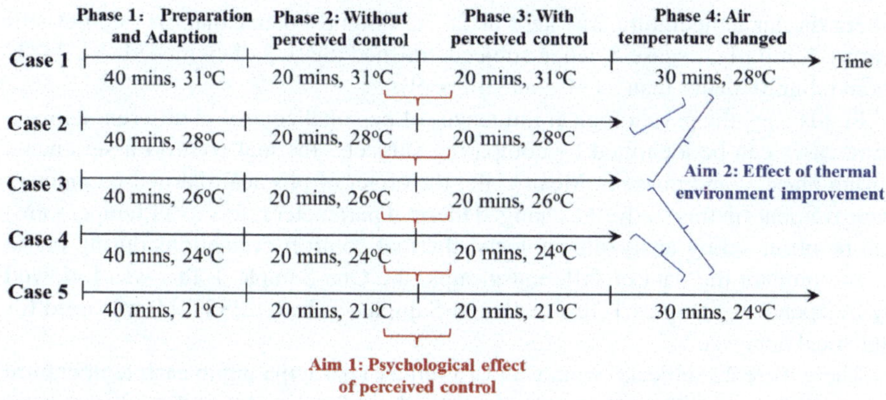

Fig. 5.9 Experimental protocol

Thermal Comfort Monitoring Station (LSI) with a precision for temperature measurement of ±0.1 °C (15–35 °C) and a precision for relative humidity measurement of ±3% (20–80%). The monitoring sensors were placed 0.5 m from the subject and 1.1 m above the floor. The room temperature outside the chamber was controlled close to the air temperature inside, which ensures the globe temperature inside the chamber be stable and be close to the air temperature. Hence, subjects' thermal comfort perception was mainly influenced by air temperature inside the chamber. Four experimental phases were designed to fulfill the objectives of this study.

Phase 1: preparation and adaptation. At the beginning of this period, subjects were supposed to be calm down, to change clothes and to learn about experimental procedures. After that, they spent 40 min to adapt themselves to approximately uniform thermal conditions. The time of 40 min was determined by a preliminary test aiming to see how long it takes a subject to achieve thermally steady status.

Phase 2: without perceived control. In phase 1 and phase 2, no space conditioning control approaches were available to subjects. Even they might feel hot or cold, they had to endure the thermal condition until phase 2 finished. It should be noted that all subjects were informed that there might exist some uncomfortable cases before they agreed to participate in this study. To guarantee the subjects' health, the five temperature cases were not extreme conditions. If a subject really felt intolerable during any test period, he/she was allowed to quit.

Phase 3: with perceived control. At the beginning of this phase, subjects were given a bell and were informed that cooling or heating would be available soon, which was a positive psychological hint. They were informed that they could declare their discomfort complaints by ringing the bell (no limited ringing times), then the experimenter was supposed to adjust the indoor temperature to a comfortable level. But in fact, the physical thermal environment was not really changed. It would be maintained as same as phase 2 during this 20 min.

Phase 4: air temperature changed. This phase was only conducted in case 1 and case 5. In case 1, once the subjects finished phase 3, they moved into the other room, where the air temperature was kept at 28 °C (slightly better than 31 °C, but still warm). Similarly, in case 5, when subjects finished phase 3, they moved to a 24 °C room (slightly better than 21 °C, but still cool).

In this way, the psychological influences of personal control in different temperature cases can be identified by comparing subjects' thermal comfort assessments during phase 2 and phase 3. Meanwhile, the effect of physical thermal parameters improvement (in this study, the changed thermal parameter refers to air temperature) can be observed by analyzing subjects' thermal comfort evaluations during phase 4. To compare the data of different groups, the One-Sample T-Test was followed by Independent Samples T-Test or Paired-Samples T-Test. SPSS 19.0 was used for statistical analysis.

There were 22 subjects (9 male and 13 female) participating in each temperature case, and the same 22 participants attended all the 5 cases. All of the subjects were young and healthy college students (freshmen and sophomores) who had little knowledge about thermal comfort. Before each experiment, they were required to be calm and not to do strenuous activities. They were required to wear uniform clothes prepared in advance: thin short-sleeved shirts, thin trousers, and lightweight shoes (with measured clothing insulation of 0.57 clo as a total). Additionally, all participants in the climate chamber were asked to remain seated and engaged only in sedentary activities. During every single test, only one subject presented in the climate chamber and he/she was reminded by a sound signal to vote every 5 min. Subjects' evaluations were determined through questionnaires including thermal sensation vote (TSV), thermal comfort vote (TCV), thermal satisfaction and thermal preferences (See appendix D).

5.3.2 The Psychological Effect of Personal Control

Table 5.5 summarizes the measured thermal parameters and the calculated PMV values during different experimental conditions. It indicates that the thermal parameters were effectively controlled in accordance with the designed set-points.

Figure 5.10 illustrates thermal comfort and thermal satisfaction as a function of TSV. The data includes results from all experiment phases (1–4). A second-order polynomial regression function could be fitted to the data distributions, which reveals that the neutral thermal sensation (TSV = 0) corresponds to the most comfortable and satisfied assessments.

As mentioned in the experimental protocol, for each temperature case, physical thermal parameters in phase 3 were maintained at the same level as phase 2. The only difference between these two phases was whether subjects got the hint about their ability to control over indoor air temperature, which was purely a psychological impact, rather than physical factors.

Figure 5.11 illustrates the TSV distribution of phase 2 and phase 3 from the 5 temperature cases. Compared with phase 2 (without perceived control), subjects in phase 3 (with perceived control) tended to have more thermal sensation votes to

Table 5.5 Statistical summary of measured thermal parameters

Temperature case	21 (°C)		24 (°C)		26 (°C)		28 (°C)		31 (°C)	
	Mean	Std	Mean	Std	Mean	Std	Mean	Std	Mean	Std
Ta (°C)	21.1	0.15	23.9	0.1	25.9	0.11	28.1	0.12	31.1	0.13
Tg (°C)	21.1	0.1	24	0.1	25.9	0.12	27.9	0.1	30.9	0.1
v (m/s)	0.08	0.05	0.05	0.06	0.08	0.05	0.09	0.1	0.1	0.06
RH (%)	48.1	4.9	49.2	3.6	47.7	2.9	47.9	3.6	49.7	4.5
PMV	−1.63	0.06	−0.58	0.07	0.12	0.07	0.83	0.09	1.94	0.1

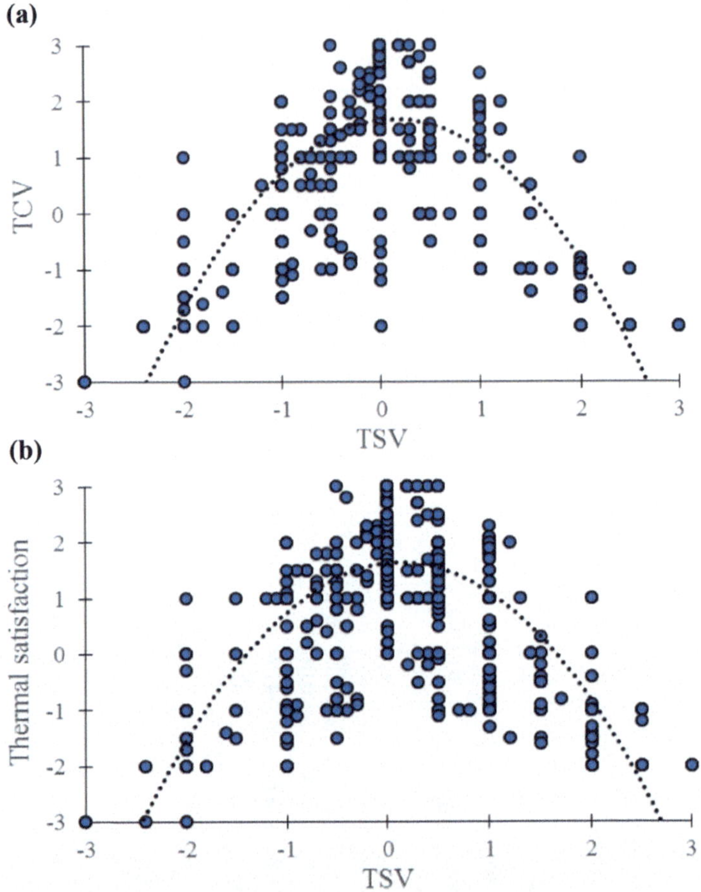

Fig. 5.10 Relationships between thermal comfort-related indices: **a** Thermal sensation and thermal comfort; **b** Thermal sensation and thermal satisfaction

"neutral" in the non-neutral temperature cases, which resulted in lower TSVs in warm conditions and higher TSVs in the cool side. Figure 5.12 shows if subjects were provided a positive psychological hint in phase 3, their thermal discomforts would be alleviated and they tended to express better comfort assessments.

To verify whether psychological impact of perceived control on thermal comfort assessment was significant, independent samples t-tests were conducted on TSVs and TCVs between phase 2 and phase 3. The test results are listed in Table 5.6, where the significance of the difference is labeled as 'strong' for p-value < 0.05. The differences between TSVs and TCVs between the two phases are significant in both warm and cool sides.

Figure 5.13 compares the regression lines of TSVs from both conditions (with personal control vs. without perceived control). All the data dots are the average

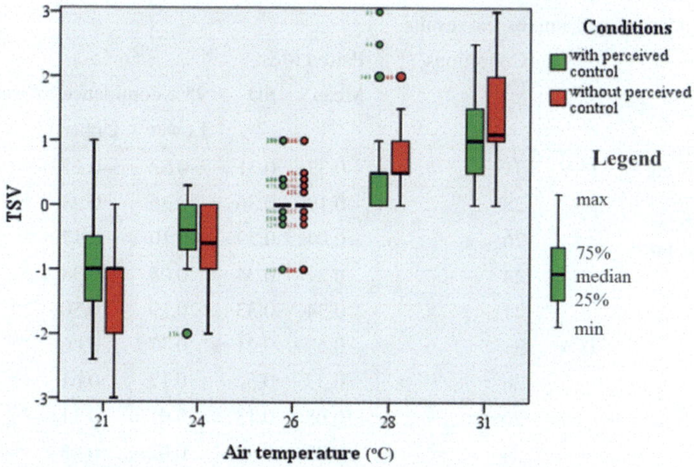

Fig. 5.11 Effect of perceived control on thermal sensation

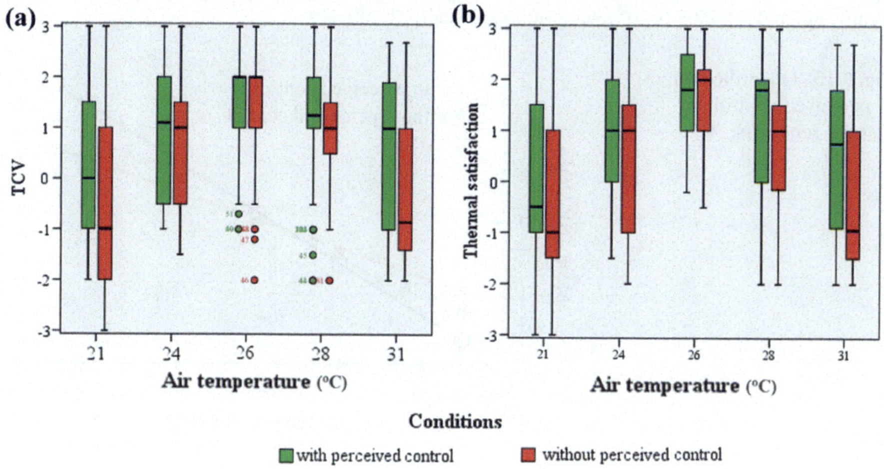

Fig. 5.12 Effect of perceived control on comfort assessment: **a** Thermal comfort; **b** Thermal satisfaction

values for all subjects in each temperature case and each experimental phase. The positive effect of perceived control can be observed from the "scissors difference", which is revealed more significant in severer temperature cases (21 and 31 °C). It is also inferred from the figure that the psychological effect is part of the reason which causes deviations between TSVs and PMV values. Similarly, Fig. 5.14 compares the regression lines of thermal comfort assessments. It shows the perceived control can alleviate subjects' thermal discomfort, especially in severe conditions such as 21 and 31 °C.

Table 5.6 Independent samples test results

		Conditions (°C)	Paired t-test				
			Mean	SD	95% confidence		Significance
					Lower	Upper	
With perceived control (phase 3) versus Without perceived control (phase 2)	TSV	31	−0.37	0.31	−0.52	−0.22	**
		28	−0.19	0.36	−0.36	−0.24	*
		26	0.04	0.29	−0.10	0.17	o
		24	0.21	0.34	0.05	0.38	*
		21	0.34	0.33	0.19	0.50	**
	TCV	31	0.57	0.21	0.47	0.66	**
		28	0.37	0.52	0.12	0.61	**
		26	0.05	0.13	−0.01	0.11	o
		24	0.25	0.48	0.02	0.48	*
		21	0.54	0.30	0.40	0.69	**

Note The significance test outcomes were interpreted as follows: $p \leq 0.001$ or "***" means highly significant, $0.001 < p \leq 0.01$ or "**" means significant, $0.01 < p \leq 0.05$ or "*" means weakly significant, and $p > 0.05$ or 'o' means not statistically significant

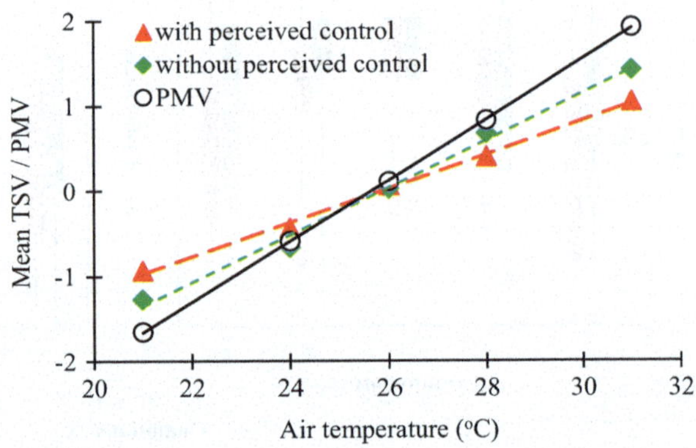

Fig. 5.13 Quantified effect of perceived control on thermal sensation

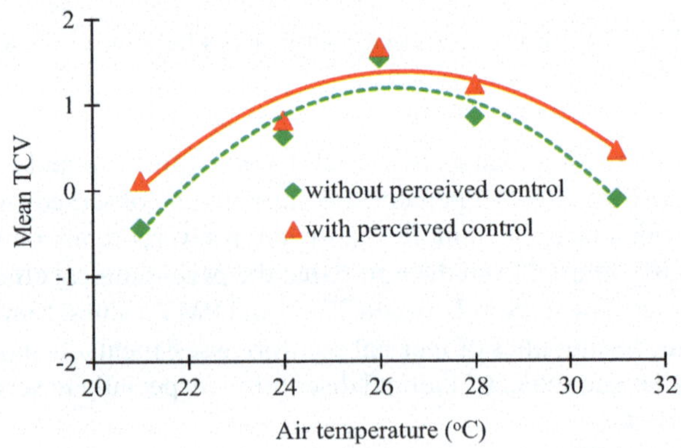

Fig. 5.14 Quantified effect of perceived control on thermal comfort

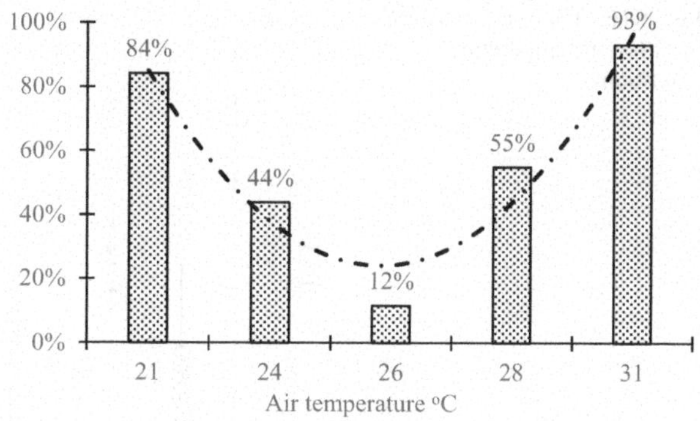

Fig. 5.15 Percentage of subjects who want to improve their thermal environment

What is the underlying psychological reasons between personal control and thermal comfort? Nikolopolou [14] once explained that people who have a high degree of control over a source of discomfort intend to be tolerant and less annoyed by the discomfort, and their negative emotional responses are greatly reduced. Moreover, it is not important whether people really use the control approach or not, the critical issue is only if the approach is available. Nicol et al. [15] also believed that people with more opportunities to adapt themselves to the environment through personal control would be less likely to suffer discomfort. These opinions, which can be interpreted as that perceived control can contribute to high thermal satisfaction, is in good accordance with the core idea of the adaptive comfort theory from the psychological view.

Figure 5.15 presents the percentage of subjects in different temperature cases who want to improve their current thermal environment. An apparent parabolic regression can be observed. When temperatures departed from neutral, subjects in increasing numbers expressed their expectations by voting 'warmer' or 'cooler'. The severer thermal conditions were, the more subjects wanted to use personal control approaches. This might be the reason why perceived control has a stronger effect in severer thermal conditions.

5.3.3 The Physical Effect of Persona Control

As described in the experimental protocol, the slightly changed air temperatures in phase 4 were designed to simulate the effectiveness when personal control approaches were utilized. Figure 5.16 shows the TSVs of different temperature cases. Most subjects reported thermally neutral when they moved from a hot room to a warm one (from 31 to 28 °C case) or from a cold room to a cool one (from 21 to 24 °C case). Experiencing this positive change of thermal conditions, their TSVs were even much better than those kept staying in 28 or 24 °C, and seemed as they entered a thermally neutral environment (26 °C).

Fig. 5.16 Thermal sensation in different temperature conditions

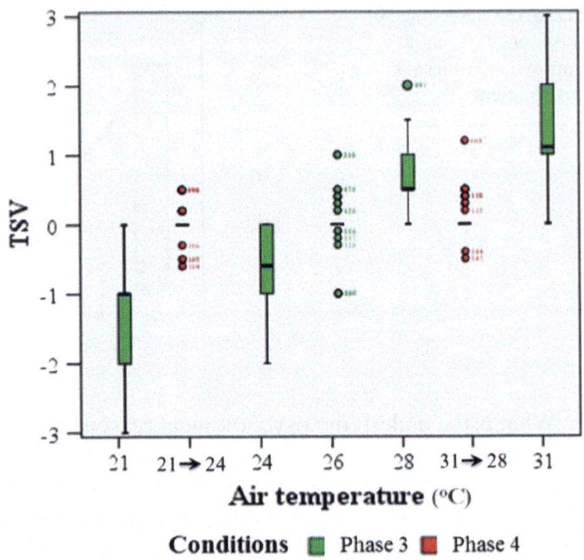

Figure 5.17 illustrates comfort-related evaluations in different temperature cases. The results indicate the subjects' comfort evaluations were greatly improved when they moved from a poor thermal environment to a slightly better one, despite the fact that they were still in a relatively poor condition. The TCV and thermal satisfaction of phase 4 were even higher than those kept staying in the neutral thermal environment (26 °C).

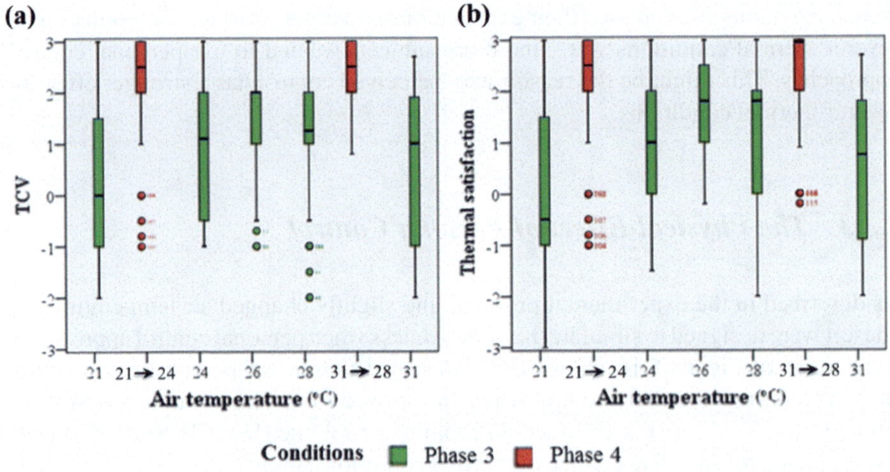

Fig. 5.17 Comfort assessment in different temperatures: **a** Thermal comfort; **b** Thermal satisfaction

Figures 5.16 and 5.17 show subjects' thermal discomfort were significantly reduced through a slight improvement of air temperature in phase 4. But, what is the reason behind this phenomenon? Fanger [16], in his explanation why PMV model overestimated the sensation of warmth in non-air-conditioned buildings, once introduced the concept of "*expectation*" as the 7th parameter in the heat balance model of thermal comfort. The expectation theory believes that people living in poor thermal environments for a long time may think it is their destiny, and they would judge a given poor environment as less unacceptable than those who are used to good thermal environments might think. According to the Fanger's theory, subjects in our present study might have lower expectation before they experience phase 4. Thus, a slight temperature improvement in phase 4 could meet their thermal expectations.

But a further question comes out: why their comfort perceptions could be even better than those kept staying in neutral thermal conditions? de Dear [17, 18]suggested that non-uniform and unsteady thermal environments may, in fact, provide occupants with better thermal comfort, perhaps even thermal pleasure, compared with homogenous, isothermal, and steady environments. The alliesthesia theory of de Dear stated that any external thermal stimulus which prospect of restoring body core temperature might have would be perceived as pleasant. In our present study, when subjects moved from a poor thermal environment to a slightly better one, they received a stimulus of air temperature improvement, so that their thermal pleasure perception might be aroused. More evidence between personal environmental control systems and the alliesthesia phenomenon have recently been illustrated in Ref. [9]. However, there is still a long way to go before we fully understand alliesthesia phenomenon.

5.4 Conclusions

This chapter examines the influence of personal control on thermal comfort through presenting results from field investigation and psychological chamber experiment. The following findings and suggestions are noteworthy:

(1) There are two ways for personal control to affect thermal comfort. One is the perceived control, which is psychological effects. The other is the physical effects of changed thermal conditions. Especially in cases deviated from thermal neutrality, a moderate thermal environment improvement can greatly improve the subjects' thermal comfort.
(2) Subjects with perceived control tend to have more neutral thermal sensation. The effect of perceived control is more obvious in the environment deviated from thermal neutrality and can be described by the linear relationships.
(3) In practice, it is recommended that occupants should be provided sufficient opportunities to control their thermal environments through operable windows, fans, terminal controllable conditioning systems, etc. In these ways, occupants' thermal demands could be well met and the complaints about the thermal environment might be significantly reduced.

References

1. Zhang H, Arens E, Kim DE, Buchberger E, Bauman F, Huizenga C (2010) Comfort, perceived air quality, and work performance in a low-power task-ambient conditioning system. Build Environ 45:29–39
2. Zhang H, Arens E, Pasut W (2011) Air temperature thresholds for indoor comfort and perceived air quality. Build Res Inf 39:134–144
3. Karjalainen S (2012) Thermal comfort and gender: a literature review. Indoor Air 22(2):96–109
4. Indraganti M, Rao K (2010) Effect of age, gender, economic group and tenure on thermal comfort: a field study in residential buildings in bot and dry climate with seasonal variations. Energy Build 42(3):273–281
5. Luo M, Cao B, Damiens J, Lin B, Ouyang Q et al (2015) Evaluating thermal comfort in mixed-mode buildings: a field study in a subtropical climate. Build Environ 88:46–54
6. Bauman F, Carter T, Baughman A, Arens E (1998) Field study of the impact of a desktop task/ambient conditioning system in office buildings. ASHRAE Transactions 104(98):1153–1171
7. Brager G, Paliaga G, de Dear R (2004) Operable windows, personal control and occupant comofort. ASHRAE Trans 110(2):17–35
8. Langevin J, Wen J, Gurian P (2012) Relating occupant perceived control and thermal comfort: statistical analysis on the ASHRAE RP-884 database. HVAC&R Research. 18:179–194
9. Zhang H, Arens E, Zhai Y (2015) A review of the corrective power of personal comfort systems in non-neutral ambient environments. Build Environ 91:15–41
10. Boerstra A, Kulve M, Toftum J, Loomans M, Olesen B, Hensen J (2015) Comfort and performance impact of personal control over thermal environment in summer: results from a laboratory study. Build Environ 87:315–326
11. Zhou X, Zhu Y, Qin O et al (2014) Experimental study of the influence of anticipated control on human thermal sensation and thermal comfort. Indoor Air 24(2):171–177
12. Technical Committee ISO/TC 159 and Technical Committee CEN/TC 122. ISO 7730: 2005 (2006) Ergonomics of the thermal environment—analytical determination and interpretation of thermal comfort using calculation of the PMV and PPD indices and local thermal comfort criteria. European Committee for Standardization, UK, 2006
13. BS EN ISO 9920-2009 (2009) Ergonomics of the thermal environment—estimation of thermal insulation and water vapor resistance of a clothing ensemble. British Standards Institution, UK
14. Nikolopoulou M, Steemers K (2003) Thermal comfort and psychological adaptation as a guide of designing urban spaces. Energy Build 35(1):95–101
15. Nicol J, Humphreys M (2002) Adaptive thermal comfort and sustainable thermal standards for buildings. Energy Build 34(6):563–572
16. Fanger PO, Toftum J (2002) Extension of the PMV model to non-air-conditioned buildings in warm climates. Energy Build 34(6):533–536
17. de Dear R (2011) Revisiting an old hypothesis of human thermal perception: alliesthesia. Build Res Inf 39(2):108–117
18. Parkinson T, de Dear R, Candido C (2016) Thermal pleasure in built environments: different thermoregulatory zones. Build Res Inf 44(1):20–33

Chapter 6
Adaptive Heating Balance Comfort Model

6.1 Background

Human thermal comfort is influenced by both external environmental attributors and personal factors. The environmental attributors include air temperature (TA), radiant temperature (TR) [1], air movement (VEL) [2], relative humidity (RH), and so forth, while the personal factors involve metabolic rate (MET) [3], clothing insulation (CLO), maybe age, gender, and adaptation [4]. To date, many efforts have been paid to develop models quantifying how these attributors may affect building occupants' thermal comfort. Many of these models have been documented and summarized in literature reviews like [5, 6]. Here, we briefly introduce two examples that have a profound impact in the thermal comfort research field.

First, the predicted mean vote (PMV) model built by Fanger et al. [7] is based on human body heat transfer calculation. The PMV model took 4 environmental variables (TA, RH, VEL, TR) and 2 personal factors (MET and CLO) into consideration. After its appearance, PMV has been incorporated in many building codes [8] but its accuracy remains a contested topic. For example, Humphreys [9] found that the actual thermal neutral temperature was lower than the predicted by a range of 0.8–3 °C. Brager and de Dear [10] got a deviation of −3.4 to +2.1 °C between the predicted and actual neutral temperature in a discussion of 18 filed studies. Cheung et al. [11] found that the accuracy of PMV in predicting observed thermal sensation was only 34%.

Second, the adaptive thermal comfort model proposed by de Dear et al. [12] observed that the acceptable temperature range in naturally ventilated buildings (NV) is 70% wider than that of air-conditioned (AC) buildings. Similar phenomena have been reported by many researchers in various climate zones [13, 14] and many other regional or local adaptive comfort models have been developed, among which Nicol

The original version of this chapter was revised: The figure captions were revised and the author provided references were added. The correction to this chapter is available at https://doi.org/10.1007/978-981-15-1165-3_8

M. Luo, *The Dynamics and Mechanism of Human Thermal Adaptation in Building Environment*, Springer Theses, https://doi.org/10.1007/978-981-15-1165-3_6

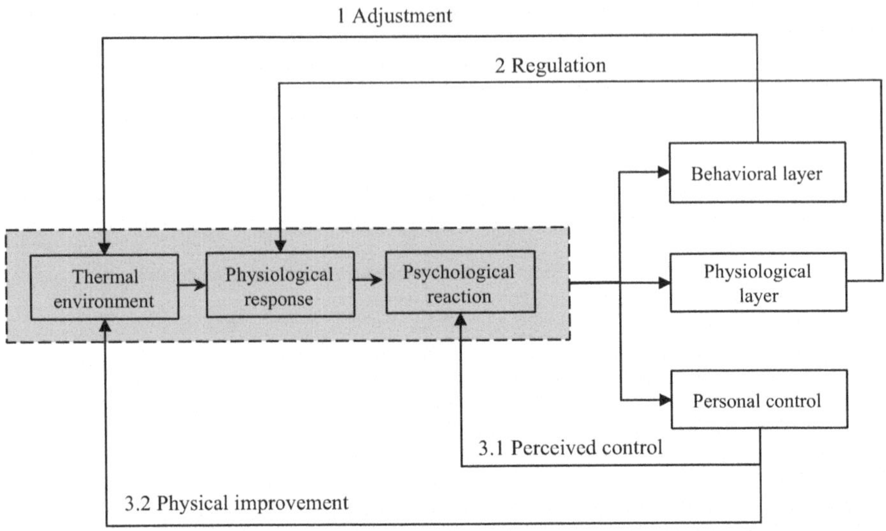

Fig. 6.1 Adaptive thermal comfort framework in buildings. *Note* The framework was modified from Fig. 1 in [23]

and Humphreys [15] derived the adaptive equation for European standard EN 15251. Notably, after a large amount of field thermal comfort studies, two publicly available databases [16, 17] have been developed, offering a solid foundation for future thermal comfort research.

However, it is worthy to note that the current PMV model and adaptive comfort model are not compatible with each other. When applying these models in real buildings, there may occur some issues. One example is that the current standards simply classify buildings into two categories (i.e., AC buildings and NV buildings) in a black-and-white way while buildings in real-world can be much more complicated, like the mix-mode buildings where co-exist both AC and NV operations [18].

Given such consideration, this chapter tries to establish a new comfort model to bridge the above two. In Chaps. 4 and 5, we have discussed the basic laws of physiological acclimation and psychological adaptation according to the theoretical framework in Fig. 6.1. Based on these findings, we are going to quantify the effects of different adaptive approaches and build a more comprehensive comfort model with a clearer mechanism.

6.1.1 Method Overview

Before developing the new model, Fig. 6.2 compares the existing two classical thermal comfort evaluation models. In the figure, the horizontal axis is outdoor temperature, and the ordinate represents indoor comfortable temperature. The black line in the figure is the indoor comfort temperature curve calculated based on the PMV model. The indoor comfortable temperature calculated by the PMV model is not

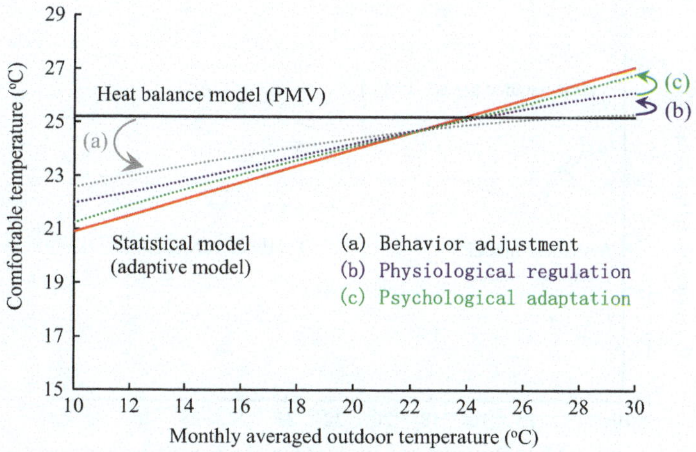

Fig. 6.2 Comparison between the adaptive comfort model and the heat balance model. *Note* The idea of embodying adaptive layers into the PMV model was inspired by [23–26] and the way to present the results was inspired by [26]

affected by outdoor temperature so that it remains constant in Fig. 6.2. The red line is the indoor comfortable temperature from the adaptive thermal comfort model, which has a linear relationship with the outdoor monthly average temperature (as shown in Eq. 1.4).

Based on the theoretical thermal adaptation framework shown in Fig. 6.1, we tried to build a link between the PMV model and the adaptive comfort model by quantitatively describing the effects of adaptation approaches such as behavioral adjustment, physiological acclimatization, and psychological adaptation (as shown by the dotted lines in Fig. 6.2).[1] We call the new model based on this concept the **Adaptive Heat Balance Comfort Model (AHBCM)**.

To date, there are several thermal comfort databases available. For example, the ASHRAE RP-884 database developed in the 1990s [17], the Smart Control And Thermal Comfort database (SCATs) developed in 2000s [19], and the more recent ASHRAE global thermal database with much larger sample size [16]. For better comparison with existing adaptive comfort models (Eq. 1.4), the RP-884 database was chosen in this study.

6.2 Adaptive Heat Balance Comfort Model Development

6.2.1 Behavior Adjustment and It's Model Embodiment

Equation 6.1 shows the basic functional relation of PMV. Where M is metabolic rate (met), t_a is air temperature (°C), t_r is radiant temperature (°C), RH is relative humidity (%), v is air velocity (m/s), and I_a is clothing insulation (clo).

[1]The dotted line in the figure is imaginary. It is used to reflect the modeling idea.

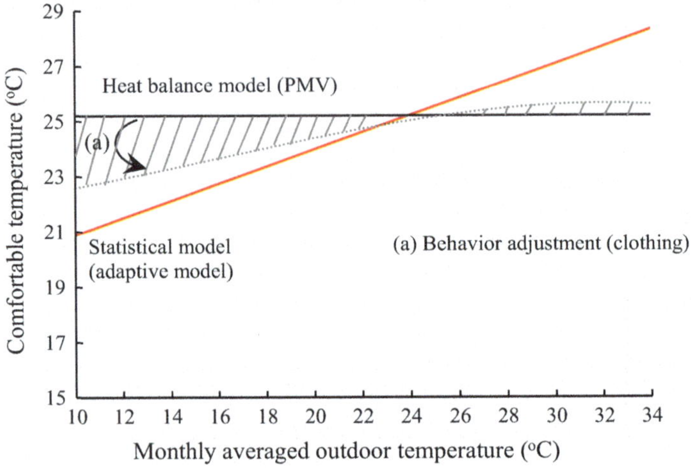

Fig. 6.3 Clothing adjustment and it's model embodiment

$$PMV_a = f(M, t_a, t_r, RH, v, I_a) \tag{6.1}$$

According to the research on RP-884 thermal comfort database [20–22], behavior adjustments in the building environment include clothing regulation, air conditioning adjustment, window closing/opening, and many other ways. Among these behaviors, clothing insulation has the greatest impact on overall thermal comfort. Therefore, this chapter defines the behavior adjustment narrowly as the changes of clothing insulation, and simply adopt the correlation between clothing insulation I_a and outdoor average temperature T_{out} described in the literature [21], as shown in Eq. 6.2:

$$I_{cl.a} = \begin{cases} 1.25 - 0.03 \cdot T_{out} & T_{out} \le 26\,°C \\ 0.45 & T_{out} > 26\,°C \end{cases} \tag{6.2}$$

Based on Eqs. 6.1 and 6.2, the relationship between the revised PMV (PMV_a) and the outdoor mean temperature T_{out} can be expressed as Eq. 6.3 (Fig. 6.3).

$$PMV_a = f(M, t_a, t_r, RH, v, T_{out}) \tag{6.3}$$

6.2.2 *Physiological Acclimation and It's Model Embodiment*

In Chap. 4, we classified physiological acclimation into two categories: long-term acclimation and short-term regulation. The long-term acclimation refers to the influence of thermal experience from a long-term view and can be described by the 'weighting factor' and 'adaptive coefficients'. The short-term regulation refers to the influence of thermal condition changes on human body thermal regulation, among which the metabolic rate changes is an example. Given that most of the existing researches on thermal adaptation do not consider the effects of long-term climate adaptation, the model construction in this chapter does not consider the acclimatization either. Instead, the metabolic rate changes in a short period can be described as Eq. 6.4, where ΔM is metabolic rate changes (met), TSV is thermal sensation vote.

$$\Delta M = 0.013TSV^2 - 0.0164TSV + 0.01 \tag{6.4}$$

From RP-884 database, we can know the relationship between thermal sensation voting and outdoor average temperature, as shown in Eq. 6.5.

$$TSV = 0.149T_{out} - 3.61 \tag{6.5}$$

In addition, it is worth noting that under cold conditions, the metabolic rate increases to raise human body heat production through chills, vasoconstriction, and so on. But under hot conditions, the metabolic rate increases to help the human body dissipate heat through perspiration, vasodilation, and so on. Based on this, a hypothesis is made here: in hot conditions, the metabolic heat production increase by one unit to help the human body dissipate two units of heat. Therefore, the changes in metabolic rate can be described by the piecewise function as shown in Eq. 6.6, in which the sedentary seating metabolic rate, which is 1.1 met, is taken as the baseline.

$$M_b = \begin{cases} 1.1 + (\Delta M) & TSV \le 0 \\ 1.1 - 2(\Delta M) & TSV > 0 \end{cases} \tag{6.6}$$

Based on Eqs. 6.5 and 6.6, the relationship between the revised PMV (PMV_b) and the outdoor mean temperature T_{out} can be expressed as Eq. 6.7.

$$PMV_b = f(T_{out}, t_a, t_r, RH, v, I_a) \tag{6.7}$$

Making PMV_b equal to zero, indoor comfort temperature curves at different monthly averaged outdoor temperatures can be calculated and shown by the purple dotted line in Fig. 6.4, where the shaded part can be regarded as the effect of short-term physiological regulation. It can be seen that the physiological regulation effect in the cold environment is stronger than that in the hot environment, which is due to the larger increase of metabolic heat production in the cold environment. However, the heat dissipation regulation of human body in hot environment has not been quantified accurately, and the assumptions put forward still need further verification.

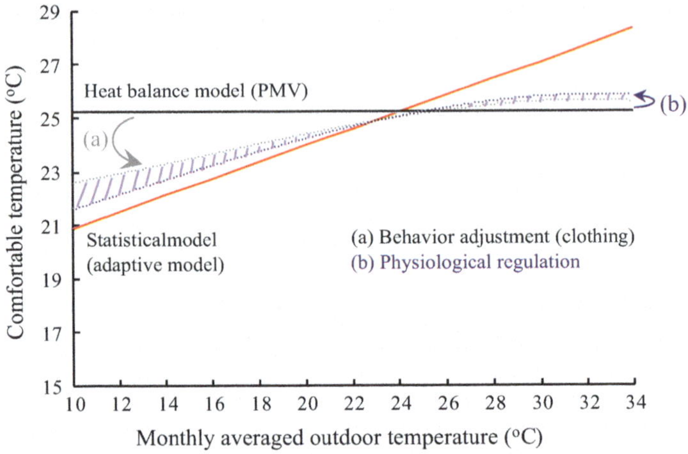

Fig. 6.4 Physiological regulation and it's model embodiment

6.2.3 *Psychological Effects and It's Model Embodiment*

In Chap. 5, the effect of perceived control on thermal comfort has been verified, and its effect has been described by linear equations. We assume the influence of perceived control is one kind of psychological adaptation and its effects (Fig. 5.14) are directly converted into the correction of PMV, as shown in Eq. 6.8. Then add and subtract to PMV, as shown in Eq. 6.9. The relationship between PMVc and monthly averaged outdoor temperature can be further described in Eq. 6.10.

$$\Delta PMV_c = 0.022T_{out} - 0.48 \quad (6.8)$$

$$PMV_c = PMV - \Delta PMV_c \quad (6.9)$$

$$PMV_c = f(M_b, t_a, t_r, RH, v, I_a, T_{out}) \quad (6.10)$$

With PMV_c equal to zero, indoor comfort temperature curves under different monthly averaged outdoor temperatures can be calculated. As shown by the dashed green line in Fig. 6.5, the shaded part can be regarded as the effect of psychological adaptation. It can be seen from the figure that the effect of psychological adaptation is more obvious in the cold or hot environment deviating from the thermal neutral temperature. However, the psychological adaptation effect shown in Fig. 6.5 depends largely on whether there are enough environment control approaches (e.g. openable windows, fans, air conditioning switchers, etc.) in actual buildings. If these personal environment controls do not exist, the psychological adaptation effect may not be as obvious as shown in Fig. 6.5.

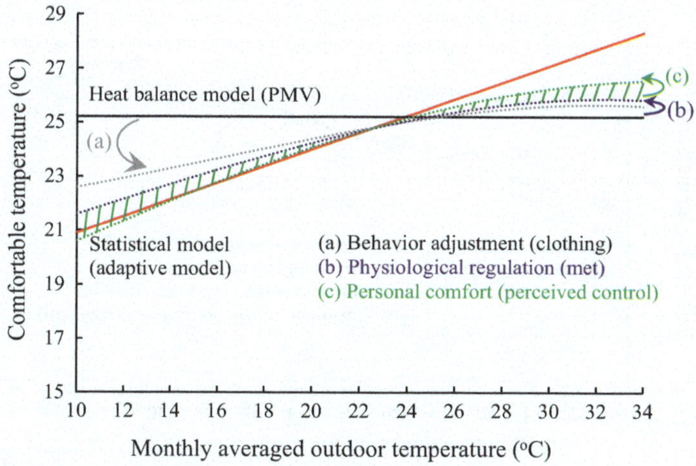

Fig. 6.5 Psychological adaptation and it's model embodiment

6.2.4 Physical Environment Improvement and It's Model Embodiment

Another adaptive dimension that should not be ignored is the changes in physical conditions. When comparing the naturally ventilated buildings and air-conditioned buildings, the former buildings have significant higher indoor air movement than the latter. Based on the statistics from RP-884 database, the air velocity in naturally ventilated buildings can be described by Eq. 6.11. By substituting it into the calculation formula of PMV, the relation can be expressed as Eq. 6.12.

$$v_d = \begin{cases} 0.02\,(m/s) & T_{out} \leq 20\,°C \\ 0.008 * e^{0.1 * T_{out}}\,(m/s) & T_{out} > 20\,°C \end{cases} \quad (6.11)$$

$$PMV_d = f(M_b, t_a, t_r, RH, v_d, I_a, \Delta PMV_c) \quad (6.12)$$

Making PMV_d equal to 0, the indoor comfortable temperatures at different monthly averaged temperatures can be calculated and shown as the orange dashed line in Fig. 6.6. The shaded part can be regarded as the effect of physical improvement. The increased air velocity in warm conditions has a significant improvement in the comfort evaluation.

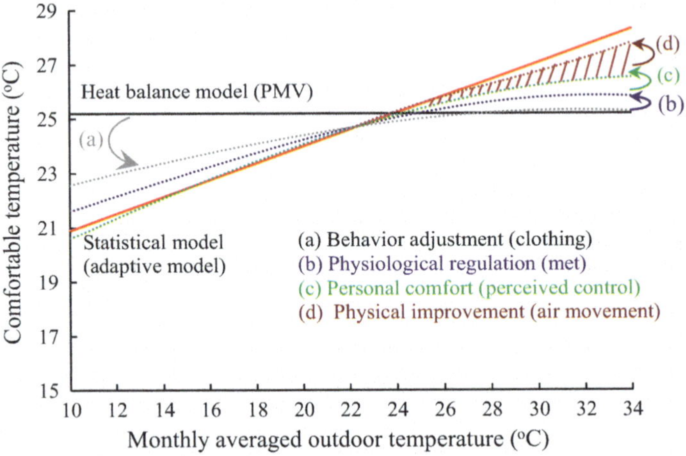

Fig. 6.6 Physical improvement and it's model embodiment

6.2.5 Effects of Demand Factor

In Chaps. 2 and 3, the 'demand factor' index was proposed to describe different subject groups' thermal comfort requirements. According to the definition of demand factor, it reflects the influence of thermal experience on occupants' thermal environment acceptance or comfort and can be directly used to modify the PMV-PPD model. If the demand factor is not considered, PMV-PPD model can be described by Eq. 6.13. If the influence of demand factor e is considered, PPD' can be described by Eq. 6.14.

$$\mathrm{PPD} = 100 - 95^{\{-(0.03353\mathrm{PMV}^4 + 0.2179\mathrm{PMV}^2)\}} \tag{6.13}$$

$$\mathrm{PPD}' = \mathrm{e} * \left[100 - 95^{\{-(0.03353\mathrm{PMV}^4 + 0.2179\mathrm{PMV}^2)\}}\right] \tag{6.14}$$

Figure 6.7 show three examples of how demand factor affects acceptable TSV range and indoor temperature range. Figure 6.7a shows that the upper and lower limits of PMV or TSV corresponding to 80% satisfaction rate for groups with different demand factors (0.8, 1, and 1.2, respectively). Figure 6.7b shows how different PMV/TSV limits would affect the acceptable temperature range. If the "demand factor" is 1, the acceptable temperature range of 80% occupants is shown as the gray area in Fig. 6.7b. If the "demand factor" is 0.8, the lower limit of 80% acceptable temperature is shown by the dashed blue line in Fig. 6.7b, and the acceptable temperature range is widened. If the "demand factor" is 1.2, the lower limit of 80% acceptable temperature is shown by the dashed red line in Fig. 6.7b, and the acceptable temperature range becomes narrower.

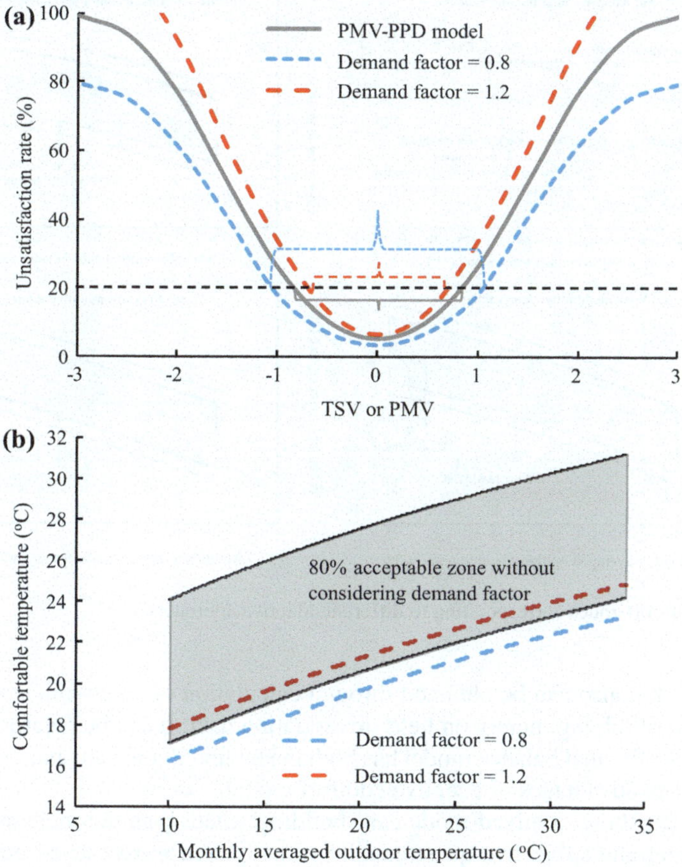

Fig. 6.7 **a** Effects of demand factor on acceptable TSV; **b** effects of demand factor on 80% acceptable temperature zone

6.2.6 *Application of Adaptive Heat Balance Comfort Model*

Figure 6.8 summarizes the adaptive heat balance model constructed above and calculates the corresponding comfort zone with the new model. Figure 6.8a is an adaptive heat balance model that comprehensively considers the four adaptive approaches of behavior regulation, physiological acclimatization, psychological adaptation, and physical parameters. The model quantifies the four adaptive approaches according to the corresponding situations, making the adaptive comfort zone much more flexible and changeable. If only physiological acclimatization and psychological adaptation are considered, the comfort zone is shown in Fig. 6.8b. If all adaptative approaches are considered, the comfort zone is shown in Fig. 6.8c. Compared with the original thermal adaptation model (Fig. 6.8d), the new model is not only flexible and

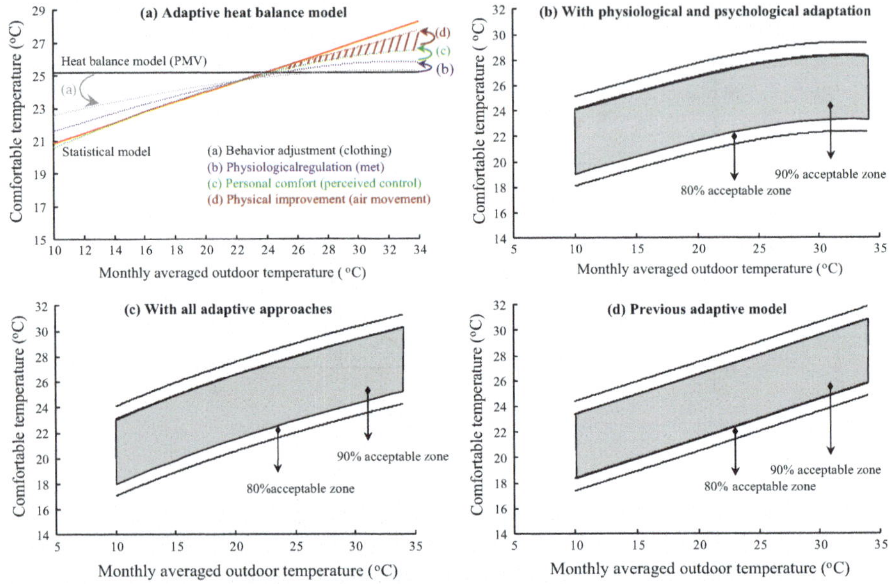

Fig. 6.8 Comfort zones corresponding to different adaptive approaches

changeable but also can be obtained through calculation of environmental parameters instead of relying merely on field investigation data. It can be regarded that the new adaptive thermal balance model has both the computational advantages of PMV and statistical advantages of adaptive comfort model.

Figure 6.9 shows a mixed-mode case building where both thermal environment measurement and subjective questionnaire investigation were carried out. Neither PMV nor existing adaptive comfort model is appropriate to evaluate its thermal comfort performance. Both two models show a large difference from the actual subjective voting results, as shown in Fig. 6.10. This is mainly due to the lack of necessary connection between PMV model and the existing thermal adaptation model, and their respective application limits are too absolute, so that the two models are ambiguous and noncommittal on the issue of whether the Mixed-mode building is applicable.

If the adaptive heat balance comfort model is adopted. It can assume that there is no adaptive approach for behavior adjustment and window opening ventilation during the air conditioning period, so the comfort zone is shown in Fig. 6.11a. As there are many adaptive ways during the natural ventilation period, so the comfort zone is shown in Fig. 6.11b. It can be found that the results from the new model are very close to the actual subjective voting results.

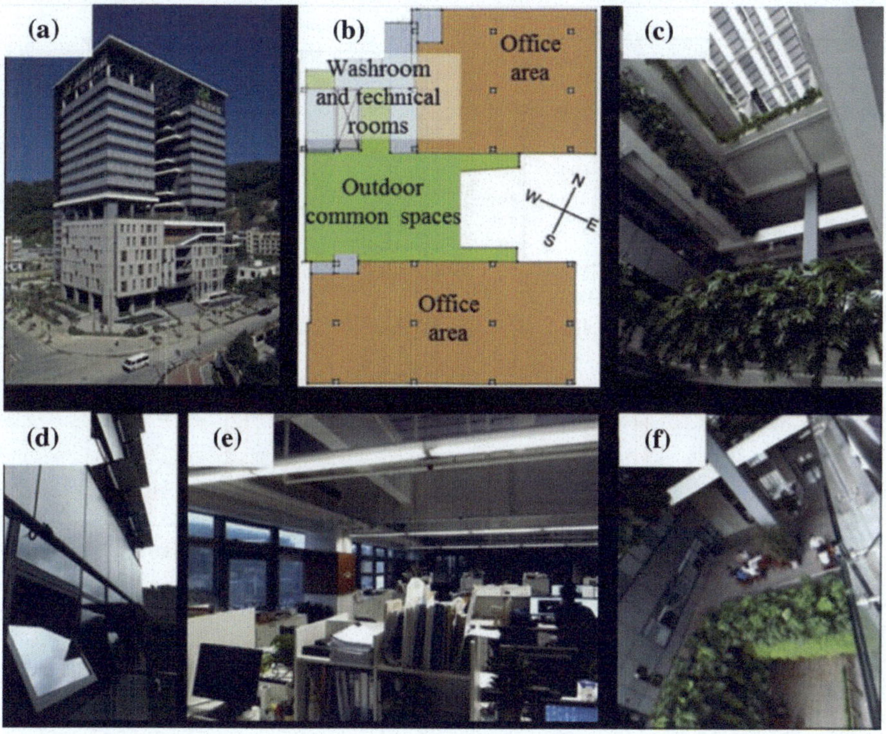

Fig. 6.9 The case building: **a** view of the building from the southeast; **b** typical floor plan; **c** outdoor corridor with vegetation; **d** operable windows; **e** typical layout of office area; and **f** outdoor platform

6.3 Conclusions

Based on the theoretical framework of thermal adaptation, this chapter combines the classical PMV model with the adaptive comfort theory framework and proposes a new adaptive heat balance comfort model. The main findings are as follows.

(1) The effects of four adaptive approaches, namely behavior adjustment, physiological acclimatization, psychological adaptation, and physical improvement, on thermal sensation and comfort temperature, are quantified separately. The results show that behavior adjustment represented by changing clothing insulation has an obvious improvement in thermal comfort evaluation in cold environment. Physical improvements represented by air movement has obviously improved comfort evaluation in hot conditions, while physiological acclimatization and psychological adaptation have obvious improving effect in conditions deviated from thermal neutrality.

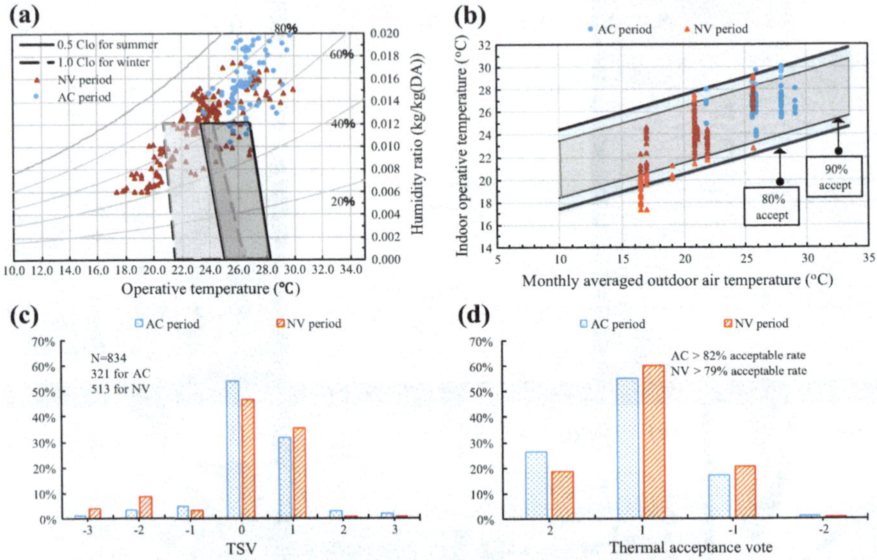

Fig. 6.10 Comfort evaluation: **a** steady-state comfort model; **b** adaptive comfort model; **c** actual thermal sensation; **d** actual thermal acceptance

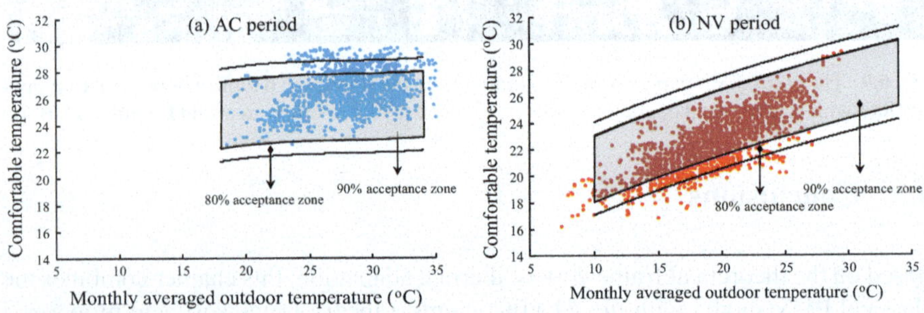

Fig. 6.11 Application of the adaptive heat balance model

(2) The "demand factor" of different populations can be reflected in the limit value of the acceptable temperature range. When the demand factor of a population is less than 1, the acceptable temperature range becomes wider. When the demand factor is higher than 1, then the acceptable temperature range becomes narrower.

(3) The adaptive heat balance comfort model has both the computational advantages of heat transfer model (PMV) and the empirical advantages of the statistical model (existing thermal adaptation model). It can be flexibly applied to the thermal comfort evaluation of actual building thermal environment.

References

1. Zhou X, Liu Y, Luo M et al (2019) Thermal comfort under radiant asymmetries of floor cooling system in 2 h and 8 h exposure durations. Energy Build 188:98–110
2. Zhu Y, Luo M, Qin O et al (2015) Dynamic characteristics and comfort assessment of airflows in indoor environments: a review. Build Environ 91:5–14
3. Luo M, Wang Z et al (2018) Human metabolic rate and thermal comfort in buildings: the problem and challenge. Build Environ 131:44–52
4. Ji W, Luo M, Cao B et al (2017) Influence of short-term thermal experience on thermal comfort evaluations: a climate chamber experiment. Build Environ 114:246–256
5. Enescu D (2017) A review of thermal comfort models and indicators for indoor environments. Renew Sustain Energy Rev 79:1353–1379
6. Zaki S, Damiati S, Rijal H, Hagishima A, Razak A (2017) Adaptive thermal comfort in university classrooms in Malaysia and Japan. Build Environ 122:294–306
7. Fanger PO (1970) Thermal comfort. Analysis and application in environment engineering. Danish Technology Press, Copenhagen
8. Technical Committee ISO/TC 159 and Technical Committee CEN/TC 122. ISO 7730 (2005) Ergonomics of the thermal environment—analytical determination and interpretation of thermal comfort using calculation of the PMV and PPD indices and local thermal comfort criteria. European Committee for Standardization, UK
9. Humphreys M (1994) Field studies and climate chamber experiments in thermal comfort research. Therm Comfort Past Present Fut 52–72
10. Brager GS, de Dear RJ (1998) Thermal adaptation in the built environment: a literature review. Energy Build 27(1):83–96
11. Cheung T, Schiavon S, Parkinson T et al (2019) Analysis of the accuracy on PMV–PPD model using the ASHRAE global thermal comfort database II. Build Environ 153:205–2017
12. de Dear R, Brager G (2001) The adaptive model of thermal comfort and energy conservation in the built environment. Int J Biometeorol 45:100–108
13. Mishra A, Ramgopal M (2013) Field studies on human thermal comfort—an overview. Build Environ 64:94–106
14. Luo M, Cao B, Damiens J, Lin B, Ouyang Q et al (2015) Evaluating thermal comfort in mixed-mode buildings: a field study in a subtropical climate. Build Environ 88:46–54
15. Nicol F, Humphreys M (2010) Derivation of the adaptive equations for thermal comfort in free-running buildings in European standard EN15251. Building and Environment 45(1):11–17
16. Foldvary V, Cheung T, Zhang H et al (2018) Development of the ASHRAE global thermal comfort database II. Build Environ 142:502–512
17. de Dear R (1998) Global database of thermal comfort field experiments. ASHRAE Trans 104:1141–1152
18. Deuble MP, de Dear RJ (2012) Mixed-mode buildings: a double standard in occupants' comfort expectation. Build Environ 54:53–60
19. McCartney KJ, Nicol JF (2002) Developing an adaptive control algorithm for Europe: results of the SCATs Project. Energy Build 34(6):623–635
20. Carlucci S, Bai L, de Dear R, Yang L (2018) Review of adaptive thermal comfort models in built environmental regulatory documents. Build Environ 137:73–89
21. Schiavon S, Lee K (2013) Dynamic predictive clothing insulation models based on outdoor air and indoor operative temperatures. Build Environ 59:250–260
22. Kingma B, Frijns A, Schellen L et al (2014) Beyond the classic thermoneutral zone including thermal comfort. Temperature 1(2):142–149
23. Zhang Y, Zhao R (2010) Literature review and discussion on human thermal adaptation in built environment. J HV&AC 40(9):38–46

24. Fanger PO, Toftum J (2001) Thermal comfort in the future-excellence and expectation. In: Proceedings on moving thermal comfort standards into 21st century, Windsor
25. Yao R, Li B, Liu J (2009) A theoretical adaptive model of thermal comfort-adaptive predicted mean vote (aPMV). Build Environ 44(10):2089–2096
26. Schweiker M, Wagner A (2015) A framework for an adaptive thermal heat balance model (ATHB). Build Environ 94:252–262

Chapter 7
Summary

Thermal comfort research aims to understand the health and comfort requirements for building occupants meanwhile reducing the energy consumption of building environment conditioning. Since the 1990s, many efforts have been paid on building occupants' thermal adaptation. But there remain some uncertain questions. First, from the phenomenon description level, the existing research mainly focused on long-term adaptation and rarely mentioned the forming process of thermal adaptation. Second, from the mechanism level, it is still unclear how the thermal adaptation was formed. Third, as for the modeling level, the current adaptive model is still "black-boxed" without explicating its mechanism behind. Given the above three flaws (or weakness), this thesis investigated building occupants' dynamic laws of thermal adaptation based on Chinese resident's perception. The comfort effects of different adaptive layers (e.g. physiological acclimatization, psychological adaptation, behavioral adjustment, and physical heat exchange) were separately examined. Based on the observations from field investigations and climate chamber experiments, a more comprehensive comfort model was developed to bridge the adaptive thermal comfort and heat balance model. The following work and conclusions are noteworthy.

(1) Through comparative online survey and field investigations in real buildings, this study examined the relationship between building occupants' indoor thermal experience and comfort perception. It has been shown that people's understandings of thermal comfort are mutually dependent on their indoor thermal experiences. Long-term comfortable thermal experiences can raise occupants' thermal expectation, while lower-grade non-neutral thermal environment can bring about thermal adaptation. The metrics of "demand factor" can be used to quantify the dynamic processes of thermal adaptation and the asymmetric trajectories of thermal comfort expectations.
(2) The physiological acclimation effects on thermal comfort have been extensively examined through three comparative climate chamber experiments. It shows

M. Luo, *The Dynamics and Mechanism of Human Thermal Adaptation in Building Environment*, Springer Theses,
https://doi.org/10.1007/978-981-15-1165-3_7

that human body physiological adaptation can be divided into short-term regulation and long-term acclimation. The former can be reflected by metabolic rate changes while the latter can be described by "weighting factor" and "adaptive coefficient" built on skin temperature and skin blood flow changes.

(3) We analyzed the influence of personal control on thermal comfort through field investigation and psychological chamber experiment. The results show that there are two ways for personal control to affect thermal comfort. One is the perceived control, which is psychological effects. The other is the physical effects of changed thermal conditions. In practice, it is recommended that occupants should be provided enough opportunities to control their thermal environments through operable windows, fans, terminal controllable conditioning systems, etc.

(4) A more explicit framework was proposed to fill the gap between the adaptive thermal comfort and the heat balance comfort model.

Based on the above work, the following findings are noteworthy:

(1) Building occupants' thermal adaptation exhibits asymmetric trajectories—it is much quicker for occupants to raise their expectations and accept a neutral indoor climate than to lower their expectations and acclimate to under-conditioned environments.

(2) The effects of physiological acclimatization and psychological adaptation can be briefly described.

(3) The new adaptive heat balance comfort model inherited both the computational advantages of heat balance model and the empirical advantages of the statistical model.

In the future, the following work can be continued:

(1) It is necessary to further understand how building occupants' thermal adaptation interact with thermal conditions changes. Especially the time-scales and dynamic laws of human thermal adaptation. The long-term thermal adaptation may take seasons and years while the medium-term adaptation may need weeks and months to accomplish. A more specific example is that the common cold symptoms as observed in Fig. 3.10 needs further investigation. More dedicated field studies and laboratory experiments should be designed to understand the missing reasons behind these observations.

(2) To build a more comprehensive comfort model, it is necessary to combine the theoretic framework of adaptive thermal comfort with big data analysis. To fulfill this purpose, it is essential to build a standardized platform for thermal comfort data collection.

(3) In addition, to improve existing averaged thermal comfort model predicting the average of a group people, applying machine learning algorithms to predict individual thermal comfort is also a promising direction [1]. Many such efforts can be found in the literature. For example, Kim et al. [2] utilized building occupants' interaction behaviors as input variables to predict individual heating or cooling demand. Li et al. [3] took the local skin temperatures to predict personal

thermal sensation. As this is a new paradigm in thermal comfort research, there remain many issues before standardizing a way to integrate machine learning algorithms with thermal comfort analysis. More comprehensive efforts on topics like comparing different algorithms, using different input variables, predicting different target thermal comfort indexes, and using different sampling methods are needed.

References

1. Kim J, Schiavon S, Brager G (2018) Personal comfort models—a new paradigm in thermal comfort for occupant-centric environmental control. Build Environ 132:114–124
2. Kim J, Zhou Y, Schiavon S et al (2018) Personal comfort models: Predicting individuals' thermal preference using occupant heating and cooling behavior and machine learning. Build Environ 129:96–106
3. Li D, Menassa C, Kamat V (2017) Personalized human comfort in indoor building environments under diverse conditioning modes. Build Environ 126:304–317

Correction to: Adaptive Heating Balance Comfort Model

Correction to:
Chapter 6 in: M. Luo, *The Dynamics and Mechanism of Human Thermal Adaptation in Building Environment*, Springer Theses, https://doi.org/10.1007/978-981-15-1165-3_6

In the original version of this chapter, figure captions were revised and the author provided references were added in Chapter 6.

The updated version of this chapter can be found at
https://doi.org/10.1007/978-981-15-1165-3_6

M. Luo, *The Dynamics and Mechanism of Human Thermal Adaptation in Building Environment*, Springer Theses,
https://doi.org/10.1007/978-981-15-1165-3_8

Appendix A
Online Survey for Indoor Thermal Comfort

Thanks for participating in the survey. We promise the information we collected is for academic purpose only. All your personal information will be kept in confidential. Thanks for your cooperation.

Background Information

1. Your gender?

○Male ○Female

2. Your age?

○<20 ○20-30 ○30-40 ○40-50 ○50-60 ○>60

Comfort assessment on your current indoor environment

3. Where do you live right now?

○An'hui	○Mecao	○Beijing	○Chongqing	○Fujian	○Gansu
○Guangdong	○Guangxi	○Guizhou	○Hainan	○Hebei	○Heilongjiang
○Henan	○Hubei	○Hunan	○Jiangsu	○Jiangxi	○Jilin
○Liaoning	○Neimenggu	○Ningxia	○Qinghai	○Shandong	○Shanghai
○Shanxi	○Shanxi	○Sichuan	○Taiwan	○Tianjin	○Hongkong
○Xinjiang	○Xizang	○Yunnan	○Zhejiang	○Oversees	

M. Luo, *The Dynamics and Mechanism of Human Thermal Adaptation in Building Environment*, Springer Theses,
https://doi.org/10.1007/978-981-15-1165-3

4. How many years have you been your current living place?

o<1year o1-2year o2-3year o3-4year o>4year

5. What is the heating method in winter in your current living place?

oRadiator heating oFloor heating oHeat pump oElectrical heater
oAlmost no heating

6. How hot/cold is the indoor temperature in your current living place?

oCold oCool oSlightly cool oNeutral
oSlightly warm oWarm oHot

7. How comfortable is the indoor temperature in your current living place?

oVery comfortable oComfortable oJust comfortable oUnclear
oJust uncomfortable oUncomfortable oVery uncomfortable

8. How humid/dry is the indoor humidity in your current living place?

oVery humid oHumid oSlightly humid oNeutral
oSlightly dry oDry oVery dry

9. How comfortable is the indoor humidity in your current living place?

oVery comfortable oComfortable oJust comfortable oUnclear
oJust uncomfortable oUncomfortable oVery uncomfortable

10. How acceptable if the indoor thermal environment in your current living place?

oVery acceptable oAcceptable oJust acceptable oUnclear
oJust unacceptable oUnacceptable oVery unacceptable

11. Do you have following symptoms in winter in your current living place?

oCommon cold oChilblain oRunning nose oInflammation
oDry lips oSore throat oElectrostatic

Comfort assessment on your hometown indoor environment

12. Where is your hometown?

○An'hui	○Mecao	○Beijing	○Chongqing	○Fujian	○Gansu
○Guangdong	○Guangxi	○Guizhou	○Hainan	○Hebei	○Heilongjiang
○Henan	○Hubei	○Hunan	○Jiangsu	○Jiangxi	○Jilin
○Liaoning	○Neimenggu	○Ningxia	○Qinghai	○Shandong	○Shanghai
○Shanxi	○Shanxi	○Sichuan	○Taiwan	○Tianjin	○Hongkong
○Xinjiang	○Xizang	○Yunnan	○Zhejiang	○Oversees	

13. What is the heating method in winter in your hometown?

○Radiator heating ○Floor heating ○Heat pump ○Electrical heater
○Almost no heating

14. How hot/cold is the indoor temperature in your hometown?

○Cold ○Cool ○Slightly cool ○Neutral
○Slightly warm ○Warm ○Hot

15. How comfortable is the indoor temperature in your hometown?

○Very comfortable ○Comfortable ○Just comfortable ○Unclear
○Just uncomfortable ○Uncomfortable ○Very uncomfortable

16. How humid/dry is the indoor humidity in your hometown?

○Very humid ○Humid ○Slightly humid ○Neutral
○Slightly dry ○Dry ○Very dry

17. How comfortable is the indoor humidity in your hometown?

○Very comfortable ○Comfortable ○Just comfortable ○Unclear
○Just uncomfortable ○Uncomfortable ○Very uncomfortable

18. How acceptable if the indoor thermal environment in your hometown?

○Very acceptable ○Acceptable ○Just acceptable ○Unclear
○Just unacceptable ○Unacceptable ○Very unacceptable

19. Do you have following symptoms in winter in your hometown?

○Common cold	○Chilblain	○Running nose	○Inflammation
○Dry lips	○Sore throat	○Electrostatic	

Appendix B
Field Investigation Questionnaire on Thermal Comfort

Thanks for participating in the survey. We promise the information we collected is for academic purpose only. All your personal information will be kept in confidential. Thanks for your cooperation.

Background information

1. Your gender?

○Male ○Female

2. Please fill your age below.

______

3. How many years you have been in college?

○<1year ○1-2year ○2-3year ○3-4year ○>4year

4. Before attending college, which province did you live?

○An'hui	○Mecao	○Beijing	○Chongqing	○Fujian	○Gansu
○Guangdong	○Guangxi	○Guizhou	○Hainan	○Hebei	○Heilongjiang
○Henan	○Hubei	○Hunan	○Jiangsu	○Jiangxi	○Jilin
○Liaoning	○Neimenggu	○Ningxia	○Qinghai	○Shandong	○Shanghai
○Shanxi	○Shanxi	○Sichuan	○Taiwan	○Tianjin	○Hongkong
○Xinjiang	○Xizang	○Yunnan	○Zhejiang	○Oversees	

M. Luo, *The Dynamics and Mechanism of Human Thermal Adaptation in Building Environment*, Springer Theses,
https://doi.org/10.1007/978-981-15-1165-3

5. What is the heating method in winter in your hometown?

○Radiator heating ○Floor heating ○Heat pump ○Electrical heater
○Almost no heating

Clothing insulation and activity level

6. Clothes you wear right now (choose all of them)

Outerwear
○Down coat ○Down vest ○Cotton coat ○Suit jacket ○Sports jacke
○Pajamas (thin) ○Pajamas (thick) ○Windbreaker (thin) ○Windbreaker (thick)
○Woolen coat

Upper wear
○Sweater (thin) ○ Sweater (thick) ○Vest ○Long sleeve shirt ○Dress

Underwear
○Long sleeve T-shirt ○ Short sleeve T-shirt ○Long Johns

Trousers
○Suit pants ○Jeans ○Pyjama trousers ○Jeans ○Cotton trousers
○Shorts ○Overskirt

Socks
○Ankle-less socks ○Ankle socks ○Knee socks ○Pantyhose

Shoes
○Leather shoes ○Sports shoes ○Sandals ○Slippers

7. What is the chair you are sitting on mostly like?

○No chair ○Wooden chair with no cushion ○Wooden chair with cushion
○Soft seat ○Others

8. What is your activity level right now?

○Sedentary sit ○Walk occasionally ○Stand ○Recline

9. What is your activity level 30 minutes ago?

○Sedentary sit ○Walk occasionally ○Stand ○Recline ○Climbing stairs
○Jogging

Comfort assessment on indoor thermal environment

10. Do you have following symptoms right now?

○Common cold ○Chilblain ○Running nose ○Inflammation

11. How hot/cold is the indoor temperature right now?

○Cold ○Cool ○Slightly cool ○Neutral
○Slightly warm ○Warm ○Hot

12. If you feel cold, what are you going to do to mitigate the cold sensation?

○Can endure, no need to take any measures ○Raise heating/air conditioning temperature
○Close the window ○Drink hot water ○Put on more clothing
○Increase activity level

13. If you feel hot, what are you going to do to mitigate the hot sensation?

○Can endure, no need to take any measures ○Lower heating/air conditioning temperature
○Open the window ○Drink cold water ○Take off more clothing
○Lower activity level

14. How comfortable is the indoor temperature right now?

○Very comfortable ○Comfortable ○Just comfortable ○Unclear
○Just uncomfortable ○Uncomfortable ○Very uncomfortable

15. How do you prefer the indoor temperature to be?

○Cooler ○No change ○Warmer

16. How humid/dry is the indoor humidity right now?

○Very humid ○Humid ○Slightly humid ○Neutral
○Slightly dry ○Dry ○Very dry

17. How comfortable is the indoor humidity right now?

○Very comfortable	○Comfortable	○Just comfortable	○Unclear
○Just uncomfortable	○Uncomfortable	○Very uncomfortable	

18. How do you prefer the indoor humidity to be?

○More humid ○No change ○Drier

19. Given the temperature and humidity, what is your overall acceptability right now?

○Very acceptable	○Acceptable	○Just acceptable	○Unclear
○Just unacceptable	○Unacceptable	○Very unacceptable	

Appendix C
Questionnaire for Physiological Test

1. What is your thermal sensation right now?

cold	cool	slightly cool	neutral	slightly warm	warm	hot
-3	-2	-1	0	+1	+2	+3

2. What is your thermal comfort right now?

Very uncomfortable	Uncomfortable	Slightly Uncomfortable	Neutral	Slightly comfortable	Comfortable	Very comfortable
-3	-2	-1	0	+1	+2	+3

3. What is your thermal satisfaction right now?

Very unsatisfied	Unsatisfied	Slightly unsatisfied	Neutral	Slightly satisfied	Satisfied	Very satisfied
-3	-2	-1	0	+1	+2	+3

4. How do you prefer the temperature to be?

○Cooler ○No change ○Warmer

5. Do you have following symptoms right now?

○Cold hand ○Cold foot ○Gooseflesh ○Chill shivering
○Muscle tension ○Others

M. Luo, *The Dynamics and Mechanism of Human Thermal Adaptation in Building Environment*, Springer Theses,
https://doi.org/10.1007/978-981-15-1165-3

Appendix D
Questionnaire for Psychological Test

1. What is your thermal sensation right now?

cold	cool	slightly cool	neutral	slightly warm	warm	hot
-3	-2	-1	0	+1	+2	+3

2. What is your thermal comfort right now?

Very uncomfortable	Uncomfortable	Slightly Uncomfortable	Neutral	Slightly comfortable	Comfortable warm	Very comfortable
-3	-2	-1	0	+1	+2	+3

3. What is your thermal satisfaction right now?

Very unsatisfied	Unsatisfied	Slightly unsatisfied	Neutral	Slightly satisfied	Satisfied	Very satisfied
-3	-2	-1	0	+1	+2	+3

4. How do you prefer the temperature to be?

○Cooler ○No change ○Warmer

M. Luo, *The Dynamics and Mechanism of Human Thermal Adaptation in Building Environment*, Springer Theses,
https://doi.org/10.1007/978-981-15-1165-3

The 7-point scale thermal sensation vote (TSV)

Scale	Meaning	Explanation
3	Hot	Perspires continuously
2	Warm	Start sweating
1	Slightly warm	Feel warm but no sweating
0	Neutral	Neither too cold nor too warm
−1	Slightly cool	Feel cool but can endure
−2	Cool	Want to put on clothing
−3	Cold	Occasionally shivering

The manufacturer's authorised representative in the EU is Springer Nature Customer Service Centre GmbH, Europaplatz 3, 69115 Heidelberg, Germany. If you have any concerns regarding our products, please contact ProductSafety@springernature.com

Printed and bound by CPI Group (UK) Ltd, Croydon, CR0 4YY
25/09/2026
02222900-0018